139
Advances in Polymer Science

Springer-Verlag Berlin Heidelberg GmbH

Mineral Fillers in Thermoplastics I

Raw Materials and Processing

Volume Editor: J. Jancar

With contributions by
E. Fekete, P.R. Hornsby, J. Jancar,
B. Pukánszky, R.N. Rothon

Springer

This series presents critical reviews of the present and future trends in polymer and biopolymer science including chemistry, physical chemistry, physics and materials science. It is addressed to all scientists at universities and in industry who wish to keep abreast of advances in the topics covered.

As a rule, contributions are specially commissioned. The editors and publishers will, however, always be pleased to receive suggestions and supplementary information. Papers are accepted for „Advances in Polymer Science" in English.

In references Advances in Polymer Science is abbreviated Adv. Polym. Sci. and is cited as a journal.

Springer WWW home page: http://www.springer.de

ISSN 0065-3195

ISBN 978-3-662-14725-2 ISBN 978-3-540-69220-1 (eBook)
DOI 10.1007/978-3-540-69220-1

Library of Congress Catalog Card Number 61642

Originally published by Springer-Verlag Berlin Heidelberg New York in 1999
Softcover reprint of the hardcover 1st edition 1999

Typesetting: Data conversion by MEDIO, Berlin
Cover: E. Kirchner, Heidelberg
SPIN: 10573681 02/3020 - 5 4 3 2 1 0 – Printed on acid-free paper

Volume Editor

Prof. Josef Jancar
School of Chemistry
Technical University Brno
Purkynova 118
61200 Brno
Czech Republic
E-mail: jancar@fch.vutbr.cz

Editorial Board

Preface

In recent years, a growing number of engineering applications of light-weight and energy efficient plastics can be found in high-quality parts vital to the functioning of entire equipments and structures. Improved mechanical properties, especially balance of stiffness and toughness, are among the most frequently desired features of the new materials. In addition, reduced flammability is considered the single most important requirement for further expansion of plastics into large volume and demanding markets such as construction and mass transport. Production of power cables also requires flame retardant cable jacketing plastics to replace or at least to reduce consumption of environmentally unsound PVC.

The two principal ways to achieve the goals mentioned above include the development of completely new thermoplastic polymers and various modifications of the existing ones. Development and commercialization of a new thermoplastic require mobilization of large human and financial resources, the latter being within the range from \$100 million to \$10 billion, in comparison to \$100 thousand to \$10 million needed to develop and commercialize polymeric material with prescribed end-use properties using physical or chemical modification of an existing plastic. In addition, the various markets utilizing thermoplastics demand large flexibility in material properties with only moderate volumes, at the best. Hence, while the majority of both commodity and engineering thermoplastics were introduced during the 18 years between 1954 and 1972, only PEEK and liquid crystalline polymers have reached the market in the last 20 years (Table 1). On the other hand, more than 4000 blends and compounds entered the marketplace between 1980 and 1997, while only very few in the years before [1].

From the arguments put forward above, one can draw the conclusion that the physical modification of the existing polymers, i.e. their compounding with solid fillers or blending with other polymers, has been the primary approach used in the development of polymeric materials „tailored“ for the majority of new engineering applications. Development and production of completely new thermoplastics with desired properties appear to be a long term, very expensive and risky alternative to the engineering of thermoplastic materials using compounding with fillers, elastomers and blending with other thermoplastics.

In addition to purely consumption related reasons, the trend towards extension of the product range by physical modifications of existing plastics is sup-

Table 1. Commercialization dates for selected thermoplastics used as matrices in compounding [1]

Year	Polymer	Producer	Abbreviation
1927	polyvinylchloride	B.F. Goodrich	PVC
1936	polyamide 6,6	DuPont	PA6,6
1938	polystyrene	Dow	PS
1939	low density polyethylene	ICI	LDPE
1954	polyurethanes	Bayer/DuPont	PUR
1954	high density polyethylene	Hoechst	HDPE
1954	polyethyleneterephthalate	ICI	PET
1956	polyamide 6	Allied	PA6
1957	polypropylene	Phillips Petroleum	PP
1958	polycarbonate	GEC/Bayer	PC
1959	linear low density polyethylene	DuPont	LLDPE
1965	polysulfone	Union Carbide/3M	PSO
1969	polybutyleneterephthalate	Celanese	PBT
1972	polyphenylene sulfide	Philips Petroleum	PPS

ported by the effort of the plastics industry to reduce the variety of produced thermoplastics and to supply the various markets with polymeric materials based on, in an ideal case, one polymer. In addition, this „unification“ also appears reasonable for greatly improving ease of recycling, as it would lead to a reduction in the need for sorting of various plastics especially in recycling of cars and durable goods (refrigerators, computer cases, batteries, etc). Due to the large variety of plastics and rubbers used in current consumer and durable goods, recycling into reusable materials has been expensive with very little commercial success. Since the advent of metallocene catalyst technology, capable of producing polyolefins with properties ranging from elastomers to hard polymers in a single process, polypropylene has been among the most frequently cited candidates to fulfil this role.

Commodity thermoplastics (polypropylene, polyethylene, polyvinylchloride, etc.) and some engineering thermoplastics (nylons, PBT, etc.) have become attractive candidates for many engineering applications replacing traditional materials and substituting for more expensive resins [2]. Relatively low price, excellent chemical resistance, good processability, potential for part consolidation and assembly simplification due to part consolidation and the possibility of modifying mechanical properties in a wide range by adding fillers and elastomers has contributed to the growing application range of filled polyolefins, especially polypropylene (PP) and polyethylene (PE). This has been made possible by creating a solid database of experimental, theoretical and empirical knowledge gathered by researchers and compounders over the last 20 years. Part of the knowledge has been published, part remains with the companies as a portion of their „know-how“.

This book aims to assemble this somewhat fragmented information, which is dispersed in a large body of literature, in a unified, coherent manner emphasizing both quantitative and qualitative relationships between structural variables and mechanical properties. The book is divided into two volumes, the first of which deals with raw materials and compounding technology. The second volume focuses on the relationships between structural variables and mechanical and physicochemical properties of particulate-filled thermoplastics. This volume deals with the current state-of-the-art in thermoplastic matrices, particulate fillers, surface modifications of fillers and with compounding and processing technologies. Despite the fact that this volume has been edited from contributions by scientists from three countries, effort has been devoted to orchestrate the individual contributions in a manner equal to monographs written by a single author.

Brno, July 1998 Josef Jancar

References

1. Gruenwald G (1992) Plastics, how structure determines properties. Hanser, Munich
2. Rubin II (ed) (1990) Handbook of plastic materials and technology. J Wiley Interscience, New York

Contents

Structure-Property Relationships in Thermoplastic Matrices

Josef Jancar

Institute of Materials Chemistry, School of Chemistry, Technical University Brno,
Purkynova 118, CZ-612 00 Brno, Czech Republic
E-mail: jancar@fch.vutbr.cz

In this contribution, structural levels existing in common thermoplastics are defined, followed by a brief discussion of the mechanisms of formation of long-chain molecules. In the next step, morphology of semicrystalline thermoplastics is discussed in considerable detail. Special paragraphs are devoted to the quantitative description of the relationship between various structural levels and elastic moduli, yield strength and fracture toughness for semicrystalline and glassy thermoplastics used as matrices for filled coumpounds. The last paragraph contains a detailed discussion of the structure-property relationships in PP.

Keywords: Thermoplastics, Polypropylene, Morphology, Elastic moduli, Yield strength, Fracture, Crystallization

Advances in Polymer Science, Vol. 139

1 Introduction

Both commercially successful products and published papers on structure-property relationships in filled thermoplastics have changed the perception of filled thermoplastics among their end-users as well as scientific and engineering communities. It has been demonstrated that thermoplastics filled with calcium carbonate, talc, mica, clay, alumina trihydrate, magnesium hydroxide, carbon black and other fillers can successfully bridge the gap in performance between neat commodity plastics, engineering resins and short fiber reinforced thermoplastics. Moreover, the commercial success of compounded thermoplastics has been indisputable in recent years, winning business contracts in such diverse markets as household items, durable goods, appliances, passenger cars, mass transport vehicles, aircraft, sporting and leisure goods, the electrical and construction industry, electronics, packaging, medical devices, etc. It has been estimated that about 30%

of polymers produced worldwide is used to produce composites and filled plastics [3]. Some sources state that more than 50% of all produced polymers is in one way or another filled with inorganic fillers to achieve the desired properties.

The original concept of compounding, i.e. to use inert, low-cost inorganic powder to replace part of the organic resin without compromising its mechanical properties and processability and, thus, to reduce the final price of the material, has been enriched substantially over the years. Inorganic fillers of specific properties are now being increasingly used to modify the physical and chemical properties of the matrix polymers, to improve processing by shortening the production cycle, to affect environmental stability of the compounds, etc. As far as the end-user's needs are concerned, an enhancement of mechanical properties of thermoplastics processed by injection molding, extrusion, blow and compression molding, is the most desired (elastic moduli, yield strength, fracture toughness, impact strength, creep, etc.). Amongst other physical properties, the desired improvement in electrical and heat conductivity, EMI shielding capability, rheological behavior and flammability is often also achieved by adding particulate fillers. Enhanced UV stability, weathering, physical cross-linking, surface activity, adsorption characteristics and barrier properties are the common physicochemical properties most affected by the fillers.

Even though the number of thermoplastics and inorganic fillers suitable for compounding is limited for various reasons, one seems to be restricted only by ones own imagination in finding new meaningful matrix/filler combinations, especially with the advent of new technologies capable of engineering the interphase region in the imminent vicinity of the filler surface, the *interphase*, which has become the center of research and development activities worldwide [272]. It is generally accepted that the interphase region in the imminent vicinity of the filler surface can greatly affect local deformation processes ahead of an advancing crack, resulting in a substantial change in the mode of fracture [4,5]. Frequently, the interphase is assumed to play the important role of a barrier to the crack propagation, of stress and energy wave reflector, of a crack driving force attenuator and elastic waves dumper [6–14]. Formation of an interphase is very strongly affected by the wetting characteristics of the filler surface and the matrix as well as the matrix rheology and compounding technology used [15]. It seems reasonable to assume that by introducing an interphase of defined properties one can control stiffness, yield strength and fracture toughness of such a "tailored" composite and, in many cases, its hydrolytic and environmental stability. Economic requirements dictate the use of an in situ deposition of the interphase in the course of a one-step melt mixing procedure. This has been accomplished by balancing the rheological forces, controlled by relative viscosities of the polymeric components and process parameters (T, rpm., time), with the thermodynamic forces with major contributions from the acid-base interactions between polymer backbone chains and polar filler surfaces in reactive compounding processes [16].

Binary combinations, i.e. filled thermoplastics or blends of two polymers, exhibit either increased stiffness or enhanced fracture resistance [17,18]. High filler loadings, above 60 wt%, used to reduce flammability or increase HDT, Vicat softening temperature or elastic moduli impart significant reduction in toughness, especially at lower temperatures or high strain rates in the course of impact loading [19]. This shortcoming partially restricts any wider use of these simple binary compounds in the construction industry, cars, mass transportation vehicles and outdoor electrical appliances.

In particulate-filled thermoplastics, the matrix is the load-bearing component and all deformation processes take place in the matrix. Particulate fillers are, in most cases, not capable of carrying any substantial portion of the load due to the absence of interfacial friction as the means of stress transfer. This is evidenced by the lack of broken particles on the surfaces of fractured filled thermoplastics. Hence, it seems appropriate to start this volume with a brief overview of the basic structural levels and manifestation of these levels in governing the mechanical properties of semicrystalline thermoplastics used in compounding.

Upon filling with an inorganic filler, various changes in the molecular and supermolecular structure of a thermoplastic may occur, depending on the filler chemical composition, surface activity, particle size and shape, filler content, compounding technique and processing as well as on the initial matrix structure. Some changes may be latent, occuring later in the service life of the plastic part as a result of service conditions. Reduction of molecular weight, crystal and spherulite size, and molecular mobility are among the most profound effects of a solid filler on the structure of matrix polymers having their reflections in the mechanical response and stability of a thermoplastic compound.

Thus, a brief survey of the current understanding of the molecular and supermolecular structures of common thermoplastics is presented first. This review starts with a brief description of the current state-of-the-art knowledge of the constitution, configuration, conformation and supermolecular structure of common glassy and semicrystalline thermoplastics. Later in this chapter, specific features of the structure-property relationships are discussed in greater detail for the most frequently filled thermoplastics. Effects of fillers on the structural variables in polypropylene, considered the most commercially important matrix, are especially emphasized.

2 Structural Levels in Common Thermoplastics

The term "structure" is a very general one, covering a broad range of possible interpretations and, thus, it needs a more exact definition in order to be clearly understood. One can define structural levels in a solid polymer as [1]: (i) constitution (atomic structure), (ii) configuration (primary bonds), (iii) conformation (spatial arrangement of the complete chain) and, in the case of semicrystalline polymers, (iv) morphology (supermolecular structure). In the following text, we will attempt to follow the origin of all the structural levels in general and, later, in the case of PP.

The long-chain character of macromolecules is the single main structural feature of this class of materials distinguishing polymers from "atomic" materials such as metals or ceramics. On the one hand, this feature, resulting in an entropy driven behavior, greatly complicates relationships between various structural parameters and physical properties of polymer solids. In the case of linear thermoplastics, the number of conformations and morphology, respectively, control the mechanical properties of glassy and semicrystalline polymers, respectively [20]. On the other hand, since monomers are relatively small molecules [1], constitution and configuration are directly reflected in the mechanical properties of cured thermosets. Crystallization of linear thermoplastics is also controlled, to a great extent, by the constitution of the main chain [21]. The entropic character of the thermomechanical behavior of polymers, compared to the energy driven behavior of metals, brings about strongly manifested viscoelastic effects and great sensitivity of physical properties of polymers to moderate changes in external conditions (temperature, pressure, time, deformation rate, etc.) [22,23]. The presence of a supermolecular structure, such as crystallites and spherulites in semicrystalline polymers, obscures correlations between the lower levels of the molecular structure and the resulting mechanical response of polymeric solids. On the other hand, the supermolecular structure is, to a vast extent, controlled by constitution, configuration and conformation of the macromolecules forming the solid polymer body.

Therefore, it is important to follow the genesis of the structural hierarchy in a solid polymer from the formation of a long-chain molecule during the polymerization process (constitution, configuration) to the structure of a solid polymeric body (conformation, morphology) in the final product in order to identify the primary structural variables which affect the mechanical properties of plastics the most profoundly. This is the first step towards controlling these properties, engineering them and tailoring polymeric materials regarding the needs of individual products, applications, design and processing techniques.

2.1 Chemical Constitution and Configuration of Common Thermoplastics

The majority of commodity thermoplastics used as matrices in compounding with fillers (PE, PP, PS) are pure hydrocarbons or halogen-substituted hydrocarbons (PVC). Their backbone chains are formed by covalent C–C bonds, which, in the simplest configuration $-CH_2-CH_2-$ (HDPE), have bond angles of 109° 28', a length of 0.154 nm and energy about 80 kcal/mole resulting in a simple "zig-zag" planar configuration of the PE molecule. In PP, the presence of a larger side group ($-CH_3$) on the asymmetric carbon atom results in steric hindrance in the spatial packing of segments leading to a helix configuration of the PP backbone chain (Fig. 1).

The second most common bond, i.e. the C–H bond, is 0.109 nm long and has an energy of 98 kcal/mole [1]. One has to keep in mind, however, that the backbone chain constitution and configuration is important only when considering

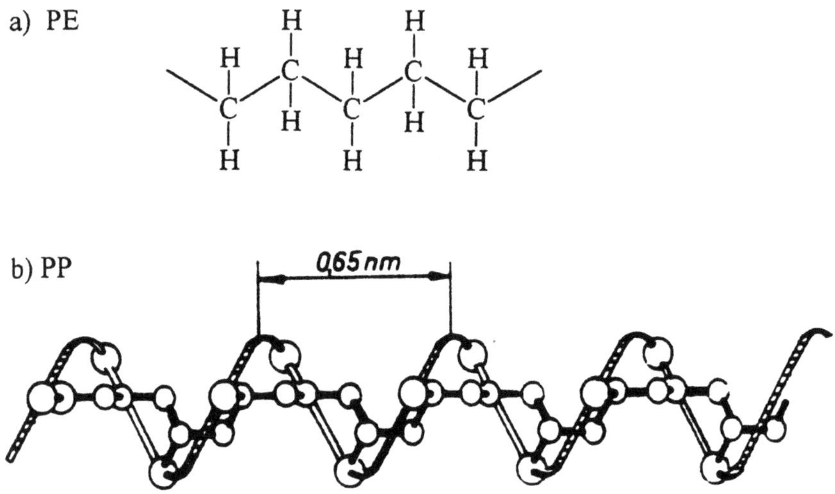

Fig. 1. Configuration of the zig-zag PE and helix PP chains

Table 1. Bond energies for bonds frequently found in polymers [1]

Bond type	Bond length (nm)	Bond energy (kJ/mole)
C–C	0.154	347–369
C=C	0.135	614–721
C≡C	0.120	811–964
C–C (aromatic)	0.139	410
C–H	0.109	414–436
C–N	0.147	293–306
C=N	0.127	614
C≡N	0.116	890
C–O	0.143	351–360
C=O	0.123	532–715
N–H	0.101	461
N…H (H-bond)	0.300	12–20
O–H	0.096	499
O…H (H-bond)	0.250	12–25
N–O	306	
C–F	0.149	432–439
Si–C	0.187	288
Si–O	0.164	368–370
B–O	500	
S–S	214	
O–S	553	

the *local behavior* of a chain, inter- and intramolecular interactions and its chemical reactivity. When considering the behavior of the whole PE molecule more than 5000 -CH_2- units long or properties of a polymer solid, the length of individual bonds is far less important than the end-to-end length of the whole chain and intermolecular interactions. Other atoms (N, O, S) also occur in conjunction with carbon in the molecular backbone chains of engineering thermoplastics (PA, PC, PPE, PPS, PET, PBT) altering their physicochemical behavior substantially compared to the pure hydrocarbon chains (Table 1).

2.2 Conformation of Thermoplastics

One of the most fascinating characteristics of macromolecules, distinguishing them from metals and ceramics, is their ability to change between various spatial arrangements of the long-chain molecule (chain conformations) very rapidly without rupturing the primary bonds (no change in configuration). In particular, chains with free rotations about the C–C bonds, so called flexible chains (PE, thermoplastic elastomers), exhibit extensive micro-Brownian motion when sufficient thermal energy is supplied, usually well above their glass transition temperature. The conformation entropy, S_c, characterizing the number of accessible conformations at given external conditions is proportional to the logarithm of the partition function of the chain:

$$S_c = k \ln\Omega \tag{1}$$

where k is the Boltzman constant and Ω is the partition function [173]. S_c is the single most important variable controlling the elastic and viscoelastic behavior of amorphous portions of glassy thermoplastics and semicrystalline polymers at temperatures above their glass transition temperature, T_g [24].

Unlike in thermosets, only weak intermolecular interactions, such as van der Waals interactions and stronger hydrogen bonds, exist between neighboring macromolecules [25] in the majority of thermoplastics, resulting in lower ultimate strength and greater propensity to creep at low or moderate loads compared to cross-linked thermosets [26]. Weak intermolecular bonds can be more or less easily broken to allow for large relative motion of individual chains [27] or, in other words, for plastic flow (Table 2). As a result, these polymers can be processed, compounded, molded, extruded and formed when heated above a critical temperature (T_g, T_m), but regain their rigid solid state upon cooling. In addition, extensive plastic deformations (cooperative changes of conformation of large molecular assemblies) result in a greater crack resistance at low temperatures, i.e. enhanced fracture toughness, and reduced elastic moduli compared to the majority of thermosets, characterized by a substantially reduced number of accessable conformations. Backbone chain stiffness, regularity of configuration and density of intermolecular cohesive energy control the glass transition temperature, T_g, in glassy polymers and melting point, T_m, in semicrystalline polymers (Table 3). Increasing the segment rigidity and/or cohesive energy density, and reducing chain mobility, increases the T_g. Similar conclusions can be drawn regarding the T_m [20].

Table 2. Molar cohesive energy of various functional groups

Group	Polymer type	Cohesive energy (kJ/mole)
$-CH_2-$	hydrocarbons	2.8
-O-	ethers	4.2
-COO-	esters	12.2
$-C_6H_{12}$	aromatics	16.4
-CONH-	amides	35.0
-OCONH-	urethanes	36.6

Table 3a. Glass transition temperature, T_g, and melting temperature for selected polymers [1]

Polymer	T_m (°C)	T_g (°C)
	Addition polymers	
LDPE	115	-120
HDPE	137	-120
PVC	175–212	87
PP	168–172	-16
PMMA	-	90–105
PS	-	95–100
PVDC	-	-15
CR	-	-45
PAN	-	105
	Condensation polymers	
Acetal	181	-85
Nylon6	225	-
Nylon6,6	265	50
Nylon12	180	140
PC	230	145
PET	255	75–80
PBT	240	-
PPO	210	

Table 3b. Properties of selected thermoplastics produced by addition polymerization

Polymer	Density (g/cm^3)	E (Gpa)	Strength (Mpa)	Elongation (%)
LDPE	0.92	0.1–0.3	8–21	50–800
HDPE	0.96	0.4–1.2	21–38	15–130
PVC	1.40	2.1–4.1	34–62	2–100
PP	0.90	1.1–1.5	28–41	10–700

Table 3c. Properties of thermoplastics with complicated backbone chain structure.

Polymer	Density (g/cm^3)	E (Gpa)	Strength (Mpa)	Elongation (%)
Nylon 6	1.14	2.8–3.4	76–83	60–300
PET	1.36	2.8–4.1	55–72	50–300
PC	1.21	2.1–2.8	62–76	110–130
PEEK	1.31	3.8	70	50–150
PPS	1.30	3.3	66	1–2

3
Formation of Long-Chain Molecules

Long-chain molecules are formed in reactions in which the elementary building blocks – monomers – are linked together by covalent bonds [28,29]. Functionality of these links controls whether a linear, branched or cross-linked macromolecule is formed. Linear and branched macromolecules consisting of a large number of repeating identical structural units are commonly termed polymers. In some cases, the reaction leading to the formation of commercially important polymers is called polymerization. Generally, three basic processes can be identified as producing high polymers. These are addition, polycondensation and ring-opening polymerizations. The type of polymerization reaction determines to a great extent the constitution and configuration of the resulting long-chain molecule and, thus, directly or indirectly the physical properties of polymeric solids (Table 4).

3.1
Addition Polymerization

The simplest and most frequently used polymerization technique is the double C=C bond polymerization [28,29]. Because of the way in which the long-chain molecules grow, this type of reaction is sometimes called either *addition* or *chain-reaction polymerization*. Generally, monomers of the general formula $H_2C{=}CRR'$ are readily polymerized after a proper initiation. The repeat unit is identical in composition to the monomer. No portion of the monomer is lost during the polymerization reaction; hence, no low molar mass species evolve. During the radical initiated polymerization of linear macromolecules, both long-chain macromolecules consisting of tens of thousands of repeat units and individual monomers exist in the reacting mixture at the same time. Most commonly, R' is the hydrogen atom and the resulting polymers are termed vinyl polymers. PE, PP, PS and PVC are the main representatives of commercially important vinyl polymers (Table 4).

The addition polymerization invariably proceeds by a chain-reaction mechanism involving three elementary steps, i.e. initiation, propagation and termination (Fig. 2). The preferred mode of monomer addition to the growing chain de-

Table 4a.

Polymer	Structure	Tensile strength (MPa)	Elongation (%)	Modulus of elasticity (GPa)	Density (Mg/cm^3)	Izod Impact Joules/,	Heat Deflection Temperature at 445 kPa	Applications
Polyethylene (PE) low density (LD) high density (HD)	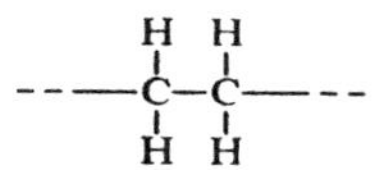	8–21 21–38	50–800 15–130	0.1–0.28 0.4–1.2	0.92 0.96	53–477 21–212	42 85	Packing films, wire insulation, squeeze bottles, tubing, household items
Polyvinyl chloride (PVC)	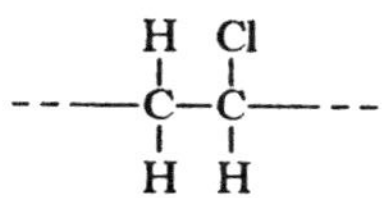	34–62	2–100	2.1–4.1	1.40			Pipe, valves, fittings, floor tile, wire insulation, vinyl automobile roofs
Polypropylene (PP)	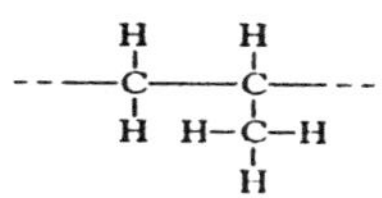	28–41	10–700	1.1–1.5	0.90	21–53	115	Tanks, carpet fibers, rope, packaging
Polystyrene (PS)	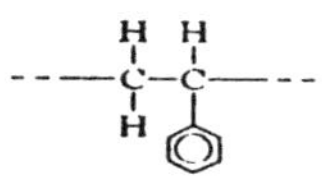	22–55	1–60	2.6–3.1	1.06	18–24	82	Packaging and insulation foams, lighting panels, appliance and furniture components, egg cartons
Polyvinylidene chloride (PVPC)	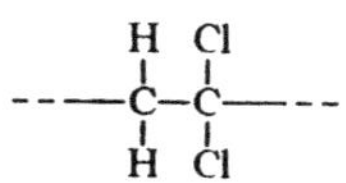	24–34	160–240	0.3–0.55	1.15	21–35	60	Packaging, pipes, draperies
Polyacrylonitrile (PAN)	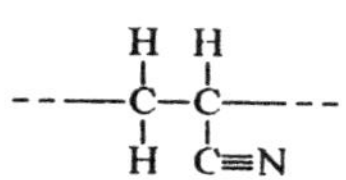	62	3–4	3.5–4.0	1.15	79.5–254	78	Textile fibers, precursor for carbon fibers, food containers

Table 4a.

Polymer	Structure	Tensile strength (MPa)	Elongation (%)	Modulus of elasticity (GPa)	Density (Mg/cm^3)	Izod Impact Joules/,	Heat Deflection Temperature at 445 kPa	Applications
Polymethyl methacrylate (PMMA) (acrylic-Plexiglas)	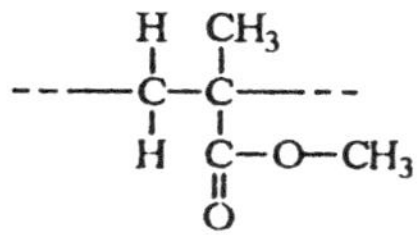	41–82	2–5	2.4–3.1	1.22	16–32	93	Windows, windshields, coatings, hard contact lenses, internally lighted signs
Polychlorotri-fluoroethylene	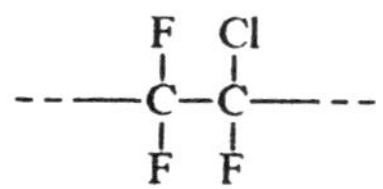	31–41	80–250	1.0–2.1	2.15	132–143	125	Valve components, gaskets, tubing electrical insulation
Polytetrafluoro-ethylene (Teflon) (PTFE)	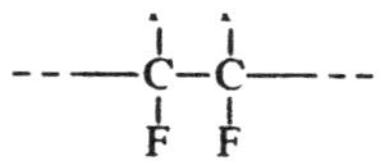	14–48	100–400	0.41–0.55	2.17	159	120	Seals, valves, nonstick coatings
Polyoxymethylene (acetal) (POM)	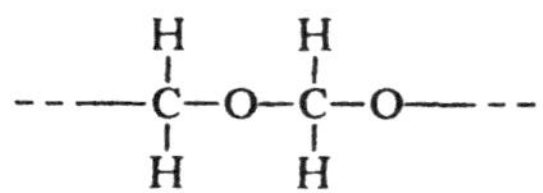	65–83	25–75	3.6	1.42	64–122	165	Plumbing fixtures, pens, bearings, gears, fan blades
Polyamide (nylon) (PA)	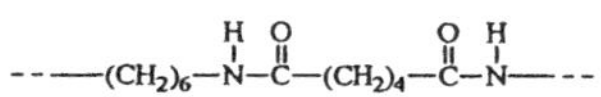	76–83	60–300	2.8–3.4	1.14	42–111	245	Bearings, gears, fibers, rope, automotive components, electrical components
Polyester (PET)	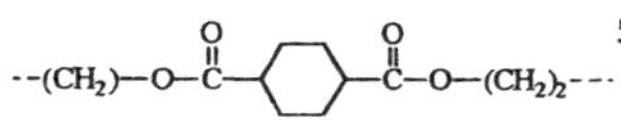	55–72	50–300	2.8–4.1	1.36	13–34	38	Fibers, photographic films, recording tape, boil-in-bag containers, beverage containers

Table 4a.

Polymer	Structure	Tensile strength (MPa)	Elongation (%)	Modulus of elasticity (GPa)	Density (Mg/cm^3)	Izod Impact Joules/,	Heat Deflection Temperature at 445 kPa	Applications
Polycarbonate (PC)	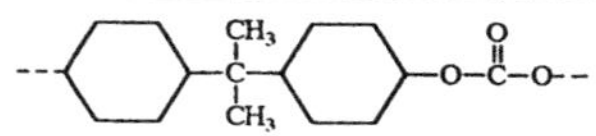	62–76	110–130	2.1–2.8	1.2	742–848	138	Electrical and appliance housings, automotive components, football helmets, returnable bottles
Cellulose	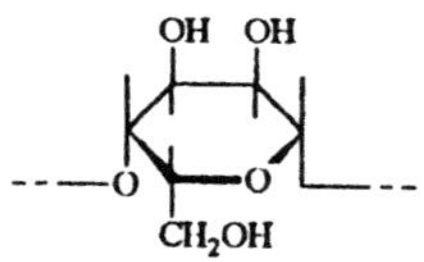	14–55	5–50	1.4–1.7	1.3	21	67	Textiles (Rayon), cellophane packaging, Ping-Ping balls, adhesives and coatings, microfilm, safety goggles
Polyimide (PI)	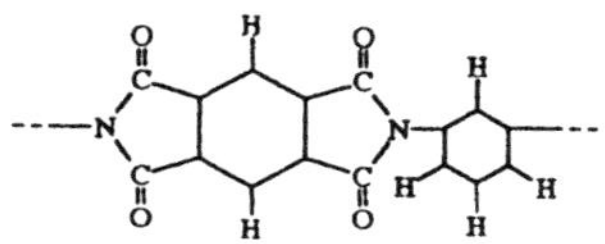	76–117	8–10	2.1	1.39	7.9	320	Adhesives, circuit boards, fibers for space shuttle
Polyphenylene oxide (PPO)	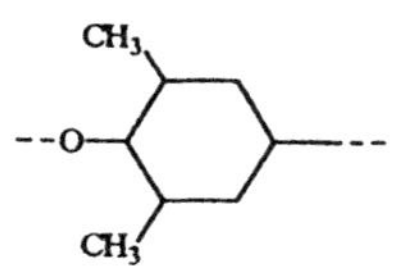	48–66	35–60	2.4	1.06–1.38	318–371	107	Automotive, business machine, and electrical components
Polyetherketone (PEEK)	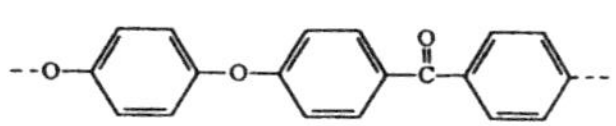	70	50–150	3.8	1.31	8.5	160	High-temperature electrical insulation and coatings

Table 4a.

Polymer	Structure	Tensile strength (MPa)	Elongation (%)	Modulus of elasticity (GPa)	Density (Mg/cm^3)	Izod Impact Joules/,	Heat Deflection Temperature at 445 kPa	Applications
Polyphenylene sulfide (PPS)	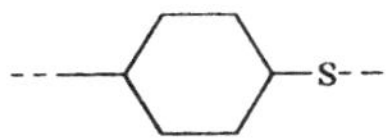	65.5	1–2	3.3	1.3	<26	135	Coatings, fluid-handling components, electronic components, hair dryer components
Polyether sulfone (PES)	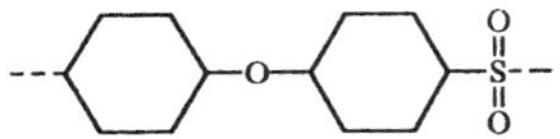	84	30–80	2.4	1.37	85	200	Electrical components, coffeemakers, hair dryers, microwave oven components
Polyacrylate	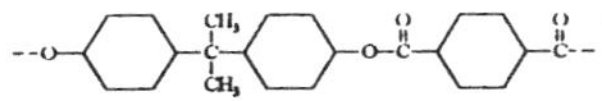	62	50–65	2.2	1.21	265	180	Trafic signals, microwave, cookware
Polyetherimide (PEI)	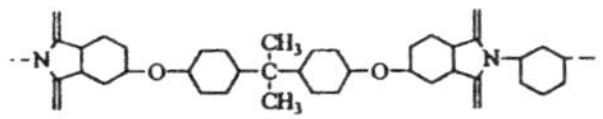	105	60	2.9	1.27	53	210	Electrical, automotive, and jet engine components

Table 4b. Thermal properties of selected thermoplastics

Polymer	Thermal linear expansivity (10^{-5} K^{-1})	Specific heat capacity (kJ kg^{-1} K^{-1})	Thermal conductivity (W m^{-1} K^{-1})
PMMA	4.5	1.39	0.19
PS	6–8	1.20	0.16
PUR	10–20	1.76	0.31
PVC	5–25	1.05	0.15–0.16
LDPE	13–20	1.90	0.35
HDPE	11–13	2.31	0.44
PP	6–10	1.93	0.24
POM	10	1.47	0.23
PA6	6	1.60	0.31
PA6,6	9	1.70	0.25
PET	10	1.01	0.14
PTFE	10	1.05	0.27
Polyethersulfone	5.5	1.12	0.18

Table 4c. Heat of combustion of selected thermoplastics

Polymer	Heat of combustion (kJg^{-1})
POM	16.9
PVC	19.9
PET	21.6
PUR	23.9
PMMA	26.2
PC	30.8
PA6, PA6,6	31.9
PS	42.2
PP	46.0
PE	46.5

pends on the substituent R, stability of the active center and steric hindrance for placing the approaching monomer to the growing chain. Generally, most monomers used in addition polymerizations are non-symmetrical resulting in three possible sequences of the monomer units in the chain (Fig. 3). These include head-to-tail and tail-to-tail and head-to-head structures termed as sequence isomerism [30]. In some cases, random grouping may also occur.

Stereoisomerism is the name for the structural variation in the position of small side groups along the long-chain molecule arising from the presence of asymmetric centers formed during polymerization. Asymmetric carbon in a propylene monomer is a typical example of such a center (Fig. 4). As a result, a regular structure of isotactic and syndiotactic polypropylene and an irregular structure of atactic PP can be produced in a single polymerization reaction.

I. primary initiation (free radical formation from organic peroxides):

$$R-O-O-R \Rightarrow 2RO\bullet$$

II. secondary initiation (growing radical):

$$R\bullet + M_1 \longrightarrow RM_1\bullet$$

$$R\bullet + M_2 \longrightarrow RM_2\bullet$$

III. propagation:

$$RM_1\bullet + M_1 \xrightarrow{k_{11}} RM_1M_1\bullet$$

$$RM_1\bullet + M_2 \xrightarrow{k_{12}} RM_1M_2\bullet$$

$$RM_2\bullet + M_1 \xrightarrow{k_{21}} RM_2M_1\bullet$$

$$RM_2\bullet + M_2 \xrightarrow{k_{22}} RM_2M_2\bullet$$

IV. termination reations:
(i) recombination

$$R(M)_x\bullet + R(M)_y\bullet \longrightarrow R(M)_{x+y}R$$

(ii) disproporcionation:

$$R-CH_2-CHX\bullet + \bullet CHY-CH_2-R \longrightarrow$$

$$\longrightarrow R-CH_2-CH_2X + CHY{=}CH_2-R$$

(iii) transfer:

$$R\bullet + XY \longrightarrow RX + Y\bullet$$

Fig. 2. Steps in addition polymerization

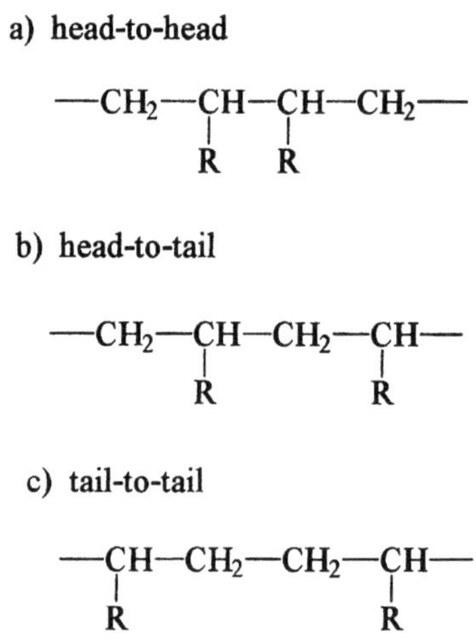

Fig. 3. Sequence isomerism in vinyl polymers

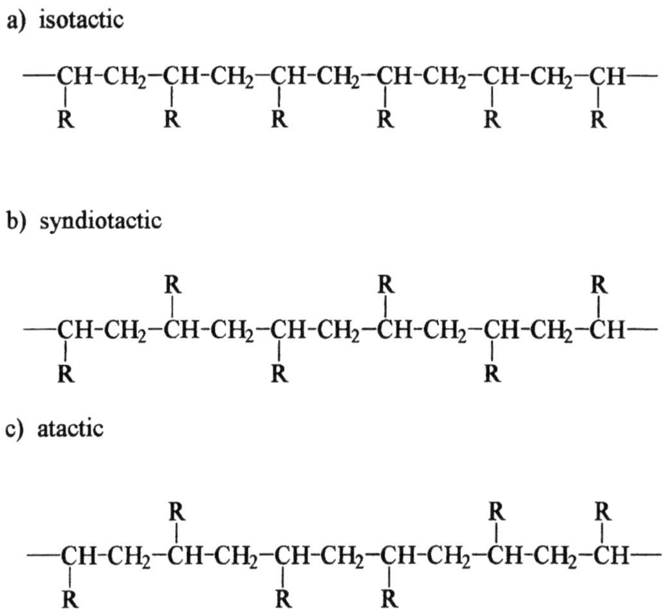

Fig. 4. Isotactic, syndiotactic and atactic isomerism in stereospecific addition polymers.

Table 5a. The most common radically polymerized commercial thermoplastics

Monomer	Chemical structure	Polymerization	Polymer
Ethylene	$CH_2=CH_2$	high pressure	PE
Vinylchloride	$CH_2=CHCl$	suspension/emulsion	PVC
Vinylacetate	$CH_2=CHOCOOCH_3$	emulsion	PVAc
Methylmethacrylate	$CH_2=CCH_3COOCH_3$	bulk	PMMA
Styrene	$CH_2=CHC_6H_5$	bulk/suspension	PS
Butadiene	$CH_2=CH-CH=CH_2$	emulsion	PBD
Chloroprene	$CH_2=CCl-CH=CH_2$	emulsion	CP

Table 5b. Commercially important polymers polymerized by ionic mechanisms

Monomer	Chemical structure	Type of reaction	Polymer
Styrene	$CH_2=CH-C_6H_5$	cation/anion	PS
Acrylonitrile	$CH_2=CH-CN$	anion	PAN
Ethylene	$CH_2=CH_2$	coordination	PE
Propylene	$CH_2=CH-CH_3$	coordination	PP

Since the chain regularity is an attribute of a development of higher levels of structural hierarchy, namely crystalline domains, stereoisomerism affects greatly the morphology and, thus, the mechanical properties of semicrystalline thermoplastics.

In addition to traditional radical initiated addition polymerizations, cation and/or anion catalyzed addition polymerizations are of great commercial importance, since PE, PP and PS are most frequently produced using this type of polymerization technique (Table 5). In addition to the vinyl monomers, vinylidene monomers, in which neither R' nor R is hydrogen, can form commercially important polymers. PMMA is a typical example of this type of thermoplastics.

3.2 Functional Group Polymerization

The functional group polymerization reaction, often termed *step reaction*, is based on the reaction of two functional groups, such as an acid and an alcohol, to form an ester or the reaction of an amine and an acid to produce an amide, etc. (Table 6). This reaction usually involves evolution of low molecular mass species, such as water, methanol, hydrogen chloride, etc. In analogy to the reactions in organic chemistry, it is, therefore, called the *polycondensation reaction*. Thermoplastic polyesters such as PET and PC are the typical and commercially most successful examples of polycondensates [31,32]. The polycondensation reaction of a diol with a diisocyanate producing polyurethane is an exception in that it does not produce a low molecular weight byproduct. Condensation polymers can be formed either by reacting a single difunctional monomer, such as a

Table 6a. Functional groups capable of polycondensation reactions

Functional Groups A	B	Polymer
OH	HOOC	polyester
OH	ClOC	polyester
Cl	NaOOC	polyester
OH	HO	polyether
COOH	HOOC	polyanhydride
OH	O=CH	polyacetal
NH2	HOOC	polyamide
SiOH	HOSi–	polysiloxane

Table 6b. Physical properties of various polyamides

Polymer	Number of $-CH_2-$ groups	Weight fraction of -CH2- in the chain	T_m (°C)	Water absorption at saturation (%)
PA6	5	0.71	225	9.5
PA11	10	0.83	190	1.9
PA12	11	0.85	180	1.5
PA6,6	6/4	0.71	265	8.5
PA6,10	6/8	0.78	225	3.3
PA6,12	6/10	0.80	210	2.8
HDPE	>5000	1.00	135	<0.01

hydroxyacid to yield a polyester, or two or more bifunctional monomers each containing the same functional groups (diacid, diol, diamine, etc.). An example is the reaction of of adipic acid with hexamethylenediamine producing Nylon 6,6 (PA 6,6) or formation of polyethylene terephthalate (PET) from ethylene glycol and terephthalic acid (Fig. 5).

Unlike the case of addition polymerization, where the backbone chain is formed from carbon atoms, in polymers formed by polycondensation, oxygen, sulfur and nitrogen atoms are frequently present in the backbone chain (Table 6). Another difference is the fact that during addition polymerization only a few sites are active and the majority of the monomer remains in its original monomeric form. In addition polymerization, high molecular weight chains co-exist with the monomer even at very low monomer conversions. On the other hand, extreme reactivity of difunctional monomers entering the polycondensation reaction leads to a fast formation of dimers, trimers and higher oligomers, so the majority of the monomer is built in bigger oligomer units very early in the reaction. The next step on the route to the high molecular weight polymer is the condensation of these oligomers into longer chains. The molecular weight increases slowly with the conversion of reactive groups and a suffi-

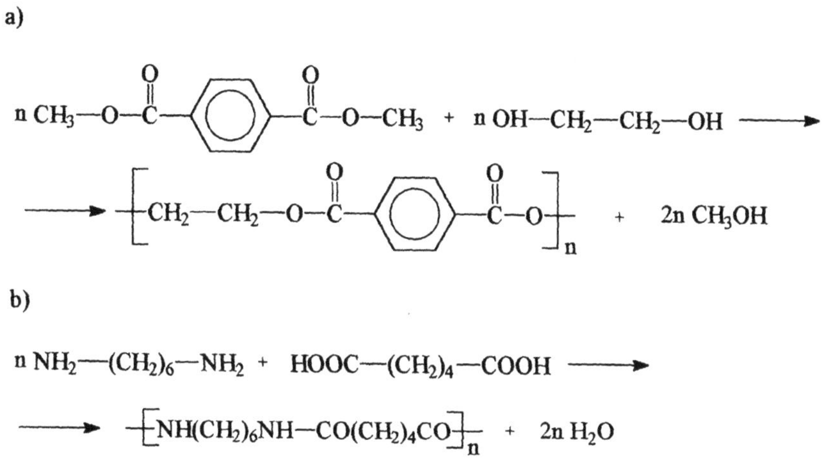

Fig. 5. Polycondensation reactions leading to typical polycondensation polymers **a** PET and **b** PA6,6

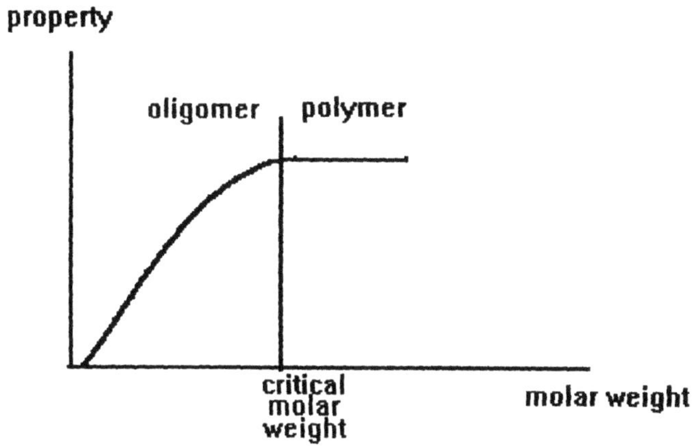

Fig. 6. Dependence of an arbitrary mechanical property on the molar weight of linear glassy thermoplastic

ciently high molecular weight polymer with commercially acceptable mechanical properties is obtained in the last steps of the polycondensation reaction or at high degrees of conversion [20]. The removal of low molecular weight byproducts has to be ensured during polycondensation in order to obtain a polymer of sufficiently high molecular weight and, thus, of usable physical properties. Generally, $M_w > 10^4$ has to be achieved in order to obtain polymers of commercially useful properties (Fig. 6).

Polymerization by a *ring-opening reaction* is confined to cyclic monomers which contain at least one heteroatom. The mechanism is very often a polyaddition-type with a product which has a polycondensation-type character. For example, ethylene oxide and other cyclic esters can be polymerized into linear chains by this type of reaction. An even more complicated example of this type of polymerization reaction is the polymerization of ε-caprolactam into Nylon 6 (PA 6).

Regardless of the polymerization technique used, the molecular weight of the macromolecule formed is strongly dependent on the conditions of polymerization (temperature, pressure, monomer purity, etc.) and it has a very important bearing on the polymer crystallinity, kinetics of crystallization, morphology, rheology, physical properties, and thermodynamic properties. Due to the nature of polymerization processes, any polymeric material almost invariably contains molecules of different lengths; hence, various averaging procedures are used for a description of such a material and polymers are characterized by various average molecular weights. The most frequently utilized averages are the number and weight average molecular weights. It can be shown that for any polymer, $M_w > M_n$, except for a monodispersity where $M_n = M_w$. Thus the M_w/M_n ratio is a good indicator of the polydispersity of a polymer [20,93]. In general, polycondensates inherently exhibit a much more narrow distribution of MW compared to addition polymers.

3.3 Regularity of Macromolecules

In many ways, the physical properties, as well as the processing characteristics, of a solid glassy polymer are controlled to a great extent by its molecular structure. In semicrystalline thermoplastics, the simple dependence is obscured by the manner in which chains interact during solidification/crystallization which occurs upon cooling and creates either long-range ordered (crystalline) or disordered (short-range order, amorphous) phases. Not only the constitution, but mostly the various ways in which the structural units can form polymer chains (configuration) and the available spatial arrangements of segments (conformation) affect the resulting properties of individual long-chain molecules as well as those of a solid polymer, respectively. Typical examples are polyethylenes and polystyrenes. HDPE and LDPE are of the same constitution but have substantially different configuration imparting variations in the amount of accessable conformations (effect of branching) and morphology. As a result, these polymers differ significantly in physical properties (Table 7). Similarly, isotactic and atactic PS have the same constitution. Their physical properties and morphology, however, differ substantially. In addition to the linear character of the backbone chain, the propensity of a thermoplastic polymer towards development of a long-range ordered supermolecular structure (crystalline phase) depends strongly on the regularity of the chain. The more regular the chain, the greater its propensity to crystallization. Isotactic PP crystalizes readily, while atactic PP

Table 7. Densities and elastic moduli of unfilled polymers

Polymer	Density (g/cm^3)	E (Gpa)
EPR	0.86	0.02
LDPE	0.91–0.93	0.1–0.3
LLDPE	0.91–0.94	0.05–0.1
HDPE	0.96–0.97	0.4–0.7
Polybutene (PB)	0.91–0.92	0.3
SBR	0.93	0.001
EVA	0.93–0.957	0.001–0.1
ABS	1.04–1.07	1.6–2.2
PS	1.05	1.6–3.3
PPO	1.06	2.5
PA6	1.12–1.13	2.4
PA6,6	1.13–1.15	2.8
PMMA	1.19	2.4–3.1
PC	1.21	2.6
Polyarylate (PAR)	1.21	2.0
Polysulfone PSO	1.24	2.5
PEEK	1.27	3.5
PEI	1.27	3.0
PET	1.34–1.39	2.2–4.2
PVC (rigid)	1.37–1.39	2.6–3.0
POM	1.41–1.43	2.8
PTFE	2.28	0.4

is an amorphous, tacky polymer with no practical use per se. Linear HDPE can attain a degree of crystallinity up to 70%. Branched LDPE has a degree of crystallinity commonly well below 20 wt%. Chains with bulky side groups or rigid chains containing aromatic rings exhibit a much weaker trend to form folded chain lamellae, the latter being able to form extended chain crystal domains such as those in the liquid crystalline polymers [33].

Because of the fact that the regularity of the formed macromolecule is in general controlled by the type of polymerization reaction, it is typically much easier to control regularity of the chain structure during a polycondenzation reaction compared to polyaddition, where the nature of propagation reaction allows variations in the mode of monomer addition to the growing chain and variability in termination reactions results in greater polydispersity of molecular weight. Principally, there are four reasons for structural variations in addition polymers: (i) structural isomerism due to formation of branches in homopolymers and variation in the monomer distribution in copolymers, (ii) sequence isomerism arising from variations in orientation of asymmetric monomer units, (iii) stereoisomerism arising from differences in configuration of asymmetric C-atoms in the chain and, finally (iv) geometric isomerism in diene polymers (*cis*, *trans*). These structural features affect the accessability of various chain conformations and, thus, are pivotal for a control of the mechanical behavior of glassy

Table 8. Effect of intensification of C–C bond clustering in thermoplastics on their elastic modulus

Polymer	Elastic modulus (Gp_a)
PE	0.2–1.0
PC	2.5
PS	3.0
PET	3.5
UHMW PE fibers	120–180
HM graphite fibers, diamond	500, 1120

polymers and for the morphology of semicrystalline polymers (supermolecular structure). Constitution and configuration govern the melting point, T_m, of the crystalline phase and the glass transition temperature, T_g, of the intervening amorphous regions, largely through controlling the main chain stiffness (Table 8).

4 Morphology of Solid Thermoplastics

The term morphology generally refers to the "supermolecular" structural level or, in other words, to the presence of a long-range order. Such a long-range translational order (symmetry) is encountered only in a crystalline phase; hence, the term morphology is used mostly in conjunction with semicrystalline thermoplastics. Polymers with the absence of any translation symmetry of the supermolecular structure are generally termed glassy (amorphous) polymers [31]. PMMA, SAN, ABS, PS, PVA, PVAc and others are commercially important glassy thermoplastics. The ability of a long-chain molecule to crystallize at all is conditional upon the sufficient regularity of the chains to permit packaging in a parallel orientation of reasonably well-ordered lattices. This implies a high degree of stereoregularity and departures from the stereoregularity, presence of co-monomers or branching can be tolerated to some extent but at the expense of poorer order and inability to form larger crystallites [21]. Semicrystalline polymers range in their properties from commodity plastics (PE, PP) to high-temperature engineering resins (PEEK) used in aerospace applications [34]. In general, properties of semicrystalline polymers are less sensitive to temperature variations, semicrystalline thermoplastics are more environmentally stable, more rigid and less permeable to low molar mass agents compared to the majority of glassy polymers [38]. In addition, while crazing is the primary deformation mechanism as a response of glassy thermoplastics to large deformations, semicrystalline polymers deform primarily plastically by shear yielding resulting in most cases in a more ductile, tougher material under given test conditions.

Some engineering thermoplastics, such as PC or PET, exist, due to their inherently high melt viscosity caused by rigid backbone chains, as supercooled liquids when solidified at moderate cooling rates, with a supermolecular structure exhibiting no long-range translation symmetry. However, both PC and PET can form crystalline phases upon long-term annealing at elevated temperature (preferentially above their T_g in order to reduce the annealing period)[35–37]. Casting thin PC and PET films from a solution can also result in the development of medium degree crystallinity in these polymers [274]. Presence of a crystalline phase greatly alters the properties of these polymers compared to their glassy analogs.

Crystalline structures, as initially produced, seldom represent even close approaches to states of thermodynamic equilibrium. Substantial changes in molecular packing within the crystalline phase occur on long term standing at temperatures close to T_m (annealing). The crystallographic symmetry of the unit cell generally remains unaltered; however, lamellae thickness (long period) may increase several times [39,40]. Compounding with fillers can result in even greater changes in the crystalline morphology of the matrix compared to the neat polymer [41].

Crystallization from quiescent melts results in textures which, although showing marked local anisotropy of chain orientation on namo- and microscales, nevertheless gives rise to properties that are essentially isotropic on a macroscopic scale [42]. Crystallization in flowing melts, such as during cavity filling in injection molding or in the film and fiber production, is a far less understood phenomenon [43,44]. During this process, the preferred molecular orientation is present already in the as-crystallized polymer greatly complicating any theoretical treatment of the process. In general, this type of crystallization under significant shear stress in the melt produces fine crystalline fibers. It commonly appears that these consist of strings of relatively small crystallites with intervening disordered layers. Formation of layers is a consequence of the considerable distortion of the molecular coils into rather elongated conformations prior to and during solidification. Molecules of high molecular weight are particularly susceptible to such an influence. Morphologies realized under extreme flow conditions are almost exclusively fibrillar whereas under more moderate conditions crystalline "shish kebabs" are formed when chain-folded crystals subsequently grow epitaxially with parallel chain orientation on the surface of fibers. These structures are common in textile fibers and extruded sheets [45,46].

The long-chain character of linear thermoplastics, resulting in folded lamellae crystalline morphology, is the reason why the degree of crystallinity in polymers is relatively low, ranging from 10 to 70%, with the exception of extended chain crystals which exist in some high-modulus synthetic fibers where crystallinity up to 95% can be attained (UHMW PE fibers Dyneema/Spectra). The way crystalline lamellae are formed in the process of crystallization by folding the chains always yields a considerable amorphous fraction on the surface of crystallites and within the interlamellae space. The spatial arrangement of the crystalline and amorphous phases controls the physical properties of a solid polymer, especially at larger strains and during fracture. The morphogenesis of the supermolecular structure in semicrystalline thermoplastics is extremely sensitive to processing conditions (undercooling, pressure, size of the produced part, etc.) and can be altered substantially by the processing method used and by additives (nucleation agents, fillers, processing aids, etc.). Substantial differences exist in the morphology of an identical polymer when processed by injection molding compared to compression molding [47].

4.1 Formation of Morphology in Semicrystalline Thermoplastics

At sufficiently high temperatures, thermoplastic polymers form quiescent amorphous viscoelastic melts, in which the chains are entangled random coils undergoing thermal micro-Brownian motion [26]. On cooling, the melt can solidify by one or both of two mechanisms: (i) crystallization (semicrystalline polymers) or, (ii) vitrification (glassy polymers). The morphology of semicrystalline polymers, frequently termed the "supermolecular structure", controls their mechanical response, environmental stability, optical properties, etc. [48]. Stereoregularity of the chain, absence of branching and stronger inter-chain interactions are some of the most important attributes of crystallizable polymers. Generally, in glassy thermoplastics, no supermolecular structure exists emphasizing the controlling role of the chemical structure of the long-chain molecules on the properties of polymeric solids.

4.1.1 *Crystallization*

Crystallization usually proceeds in two steps – nucleation and growth [21,42]. There are two principal types of nucleation – homogeneous and heterogeneous with the latter being much more common [49]. Typical homogeneous nucleus dimensions are of the order of 10^2 nm^3, while a typical polymer chain volume ranges from 10^2 to 10^4 nm^3. Thus, only a small portion of the chain is involved in forming a nucleus. There are two types of nucleus shapes – fringed micelle and folded chain [50]. The fringed micelle is thought to be a bundle of chains being parallelly oriented in a small portion of their length with the rest remaining uncrystallized (Fig. 7). Chain-folded nuclei (Fig. 8) are substantially more probable than fringed micelle nuclei in all cases when segmental mobility of the chain is relatively high.

Heterogeneous nucleation [51–55] has much greater practical importance, occuring mostly on impurities remaining from catalytic systems, on filler particles or due to the action of nucleating agents intentionally added to support nucleation. Since the presence of solid surfaces greatly reduces the free enthalpy for critical nucleation, critical nucleus size is reduced and, thus, heterogeneous nucleation and consequent crystallization can proceed at lower undercooling. The activity of latent heterogeneous nucleation centers, existing in the polymer melt but non-active under common conditions,vary and can be modified by various additives as shown for polypropylene [56]. The kinetic nucleation theory with chain folding is now a generally accepted tool for understanding the primary nucleation and isothermal growth of polymer crystals in quiescent melts [57]. The

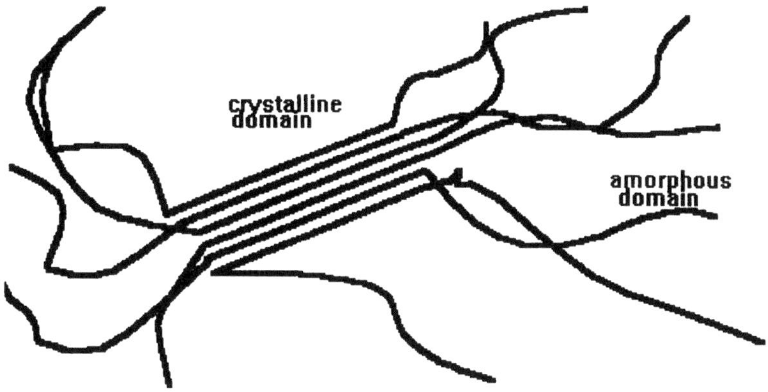

Fig. 7. Schematic representation of the fringed micelle model of crastalline polymers

state-of-the-art treatment of the chain motion and transport in the melt in the course of nucleation and crystal growth utilizes the reptation theory developed by de Gennes [58–61]. The reptation theory provides a more accurate prediction of the activation energy of chain transport compared to the classic kinetic theory and predicts the effect of molecular weight on the crystallization under different crystallization conditions [62, 63].

In the crystalline phase, the stable chain conformation has minimum potential energy. A decrease in free energy is the thermodynamic driving force of crystallization. However, the high viscosity of polymer melts leads to a diffusion controlled growth resulting in relatively low or moderate degrees of crystallinity in the majority of commercial semicrystalline polymers. Low-temperature recrystallization of solid thermoplastics occuring in relatively long time periods is one of the manifestations of the kinetic character of polymer crystallization. The kinetic character of crystallization is also documented by effects of thermal treatment or thermal history on the degree of crystallinity and the perfectness of the crystallites developed.

External conditions of crystallization such as crystallization temperature (undercooling) and pressure affect the resulting crystalline morphology substantially. Application of high pressure during crystallization can affect the intermolecular interactions, producing new crystalline forms of various thermodynamic stability [21]. As an example, polyethylene crystallization under high pressure occurs via transition of an initial hexagonal phase in the orthorhombic phase [21,38,42,64].

The early interpretation of crystallization in molten polymers was the theory of crystalline, fringed micelles [42,48,65]. Even though this theory has been proven invalid for semicrystalline polymers crystallized from dilute solutions under well-defined conditions, many processes used in forming and processing semicrystalline plastics result in the formation of a mophology similar to the idea of fringed micelles. It is clear that there is a range of ordering within the lamellae and it seems reasonable to assume that crystals grown from the melt

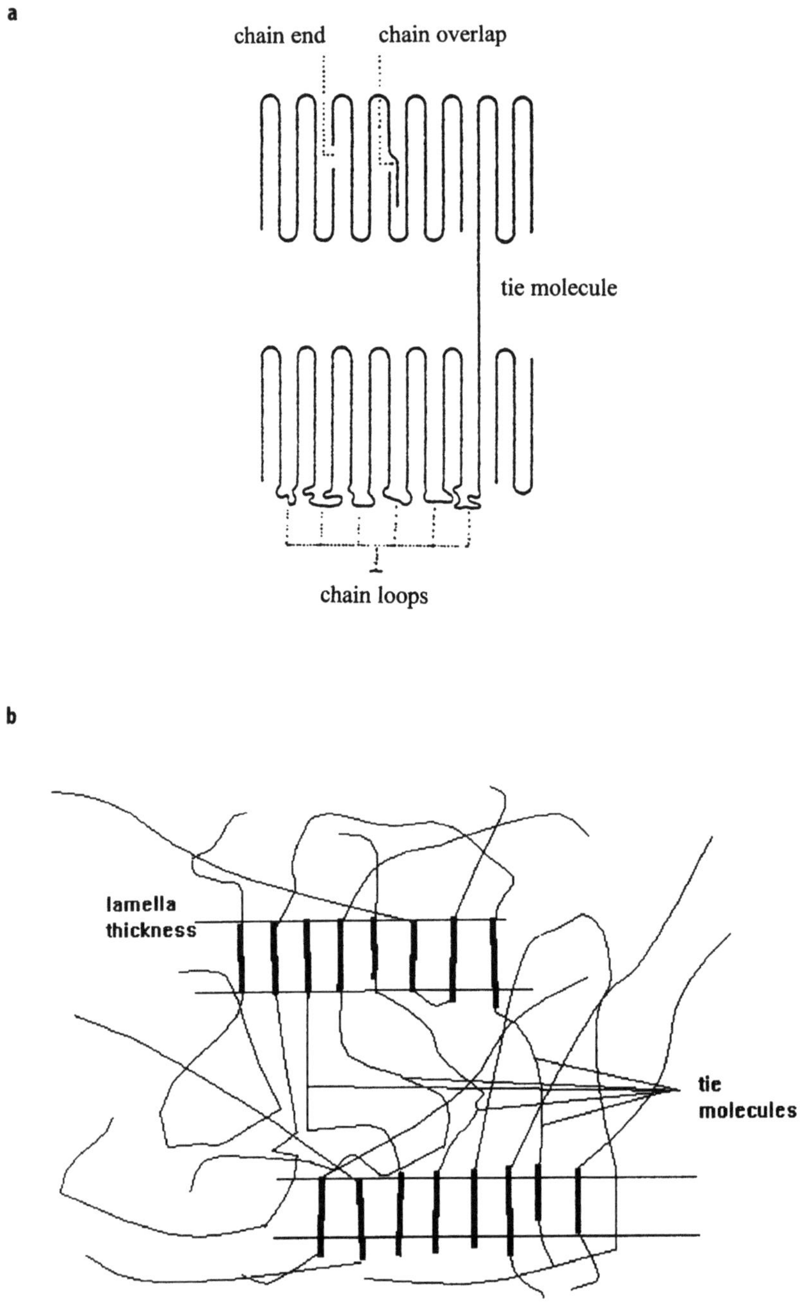

Fig. 8. Schematic representation of the folded chain model of crystalline domains in thermoplastics. **a** Lamella crystallized from dilute solution (multiple/adjacent re-entry) and **b** Lamella crystallized from melt (single entry/far distance re-entry)

formed by entangled chains are less ordered than those crystallized from dilute solutions. In particular, it is not easy to establish regularity of folding within lamella during crystallization from the melt. It is now believed that lamellae formed from dilute solutions prefer adjacent re-entry of the chain into the same lamella, resulting in many folds (10–100), whereas crystallization from the melt favors a chain to participate in several lamellae with a single entry into a given lamella or a re-entry far from the initial entry [48,57,66–76]. Typical dimensions encountered for crystalline lamellae in common PE or PP are a thickness of 10–20 nm and width of 1–10 μm [21]. These dimensions, however, vary greatly with conditions of crystallization. Polymer crystals of different polymers have one feature in common, and that is that one chain takes part in many unit cells. This is a single most important difference compared to small molecular or atomic crystals allowing the drawing of semicrystalline polymers into fibers. Unlike the case of smaller organic molecules, the surfaces of lamellae cannot be considered as cleavage planes of the polymer lattice. For cleavage to occur along these planes, primary bonds have to be broken. An extensive compilation of crystallographic parameters for synthetic polymers can be found in the work by Miller [77] and others [78]. Since there are a number of causes for irregularities in crystalline morphology, Hosemann [79] has developed a theory of disorder in polymer crystal lattices, which in its complexity goes far beyond the scope of this treatise.

4.1.2
Spherulites

The most frequently encountered aspect of polymer morphology seen in an optical microscope is spherulitic ordering, i.e. spherulites in different stages of development (Fig. 9). When crystallized from relatively quiescent melts at appreciable rates, polymers form spherulites [80]. These grow from primary nuclei associated most commonly with foreign particles and develop as globular aggregates consisting of many latch-like lamellar crystals averaging 0.1–0.5 μm in width, which fan radially outwards and branch to maintain a more or less uniformly space filling texture [81,82]. Lamellae branches in spherulites are twisted about the radial axis [65,83].

Spherulites are polycrystalline aggregates with more or less equivalent radial units [21,42,48,57]. This means that the spherulitic microstructure is a poorly coordinated array of equivalent, radiating, branching polycrystalline units tending to produce a spherical envelope. The way in which a spherulite develops depends mostly on the way it is nucleated. The molecular weight strongly affects nucleation in the sense that longer molecules usually initiate crystallization of a polymer due to their greater ability to adopt a suitable conformation. The theory of Keith and Padden [84] represents a semiquantitative model for spherulitic growth taking into account all the above-mentioned features observed experimentally. Spherulites of a polyhedral shape with a diameter ranging from 0.5 to

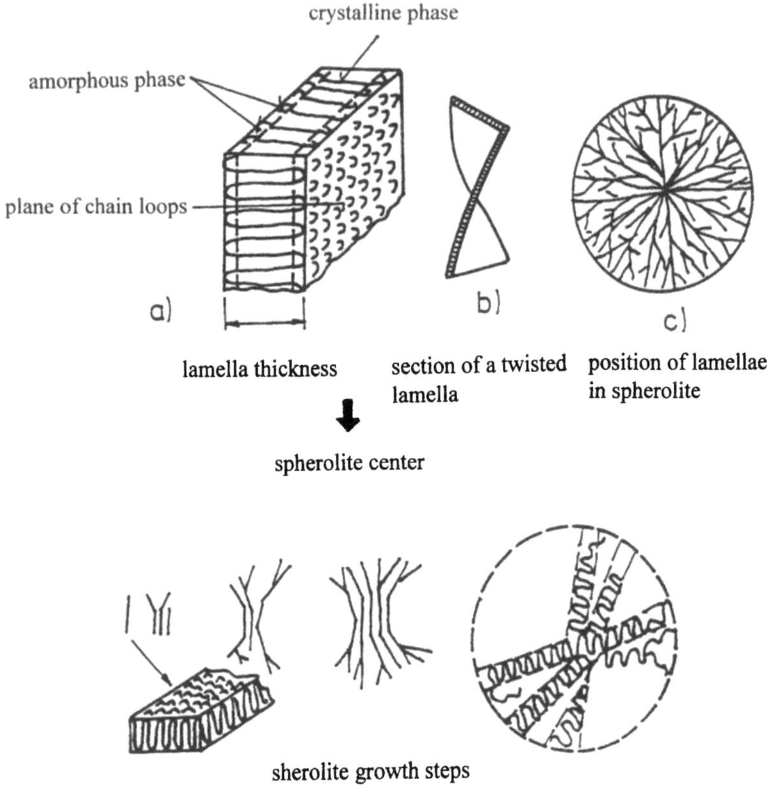

Fig. 9. Drawing of a 2D projection of a spherulite formed from radially grown lamellae with a small degree of branching

100 μm are characteristic of polymers crystallized from the melt in the absence of significant stress or flow. It is now believed that spherulites form as a consequence of certain general crystal growth conditions specific to polymers. It has to be emphasized that spherulitic texture is essentially independent of textures of lamellar units built into them.

4.2 "Supermolecular Structure" in Glassy Polymers

There is a wide range of commercially important thermoplastics that belong to the group of glassy polymers (PC, PET, PMMA, PVC, PS etc.), since in commercially important versions of these thermoplastics and, under common conditions, no long-range translational symmetry encountered in semicrystalline polymers exists [85]. As a result, these polymers are of little morphological interest; however, their commercial importance is substantial. Even though no long-range order exists, scattering studies have shown that domains of localized order may exist in amorphous polymers such as PC and PET. The published results dealing with this phenomenon are still somewhat controversial, since evidence for random coil conformation of chains in solid glassy polymers were presented [86–88] and, on the other hand, evidence supporting existence of some sort of short-range "order" in the form of nodular objects of 5 nm in size, exhibiting periodicity in electron diffraction spectra, were presented for several glassy polymers. Yeh and Geil [89,90] reported these objects of the size of 7.5 nm in amorphous PET and 12.5 nm in size in PC. These experimental observations are supported theoretically by the work of Lindenmeyer [91] predicting that the conformation of a chain in the melt or an amorphous solid is closer to irregularly folded molecules resembling folded-chain lamellae than it is to a random coil. It has been observed that upon heating at temperatures well above their Tg, crystallization occurs resulting in the development of mostly lamellar crystalline morphology greatly affecting the mechanical properties (embrittlement, increase in stiffness) and optical properties (transparency->translucency transition) of these glassy thermoplastics. Moreover, the presence of solid surfaces such as carbon and glass fibers seems to promote this type of ordering substantially [35–37]. Since some glassy polymers, such as PC and PET, can exist as both amorphous and semicrystalline solids, there has been a hypothesis proposed considering these nodular objects as crystal nuclei.

5
Structural Variables Controlling the Mechanical Properties of Thermoplastics

The mechanical properties of a polymer represent its response upon mechanical loading or straining. The vast majority of end-uses of polymers are concerned with mechanical loads and, thus, the mechanical properties of polymers are often the decisive factors for designers in selecting a suitable material. Any polymer scientist or materials engineer desires the acquisition of algorithms capable of relating structural variables to the mechanical properties, thus making the material selection more sophisticated, reliable and economically sound. As pointed out by Ward [20], any discussion of mechanical properties of solid polymers often contains macroscopic considerations as well as microscopic ones. The former result in a phenomenological description of the macroscopic mechanical response of a solid polymeric body, while the latter relates to the explanations of this behavior in terms of the chemical and physical structure of the polymer. In other words, the phenomenological description is satisfactory for the engineering/design applications and the microscopic description allows producers to develop new materials with properties tailored to the required end-use application.

One of the most profound specifics of polymers is the strong dependence of their mechanical response on the external conditions, especially temperature, strain, strain rate, duration of the loading, etc. A single polymer can show all the features of a glassy brittle solid or an elastic rubber and, eventually, viscous liq-

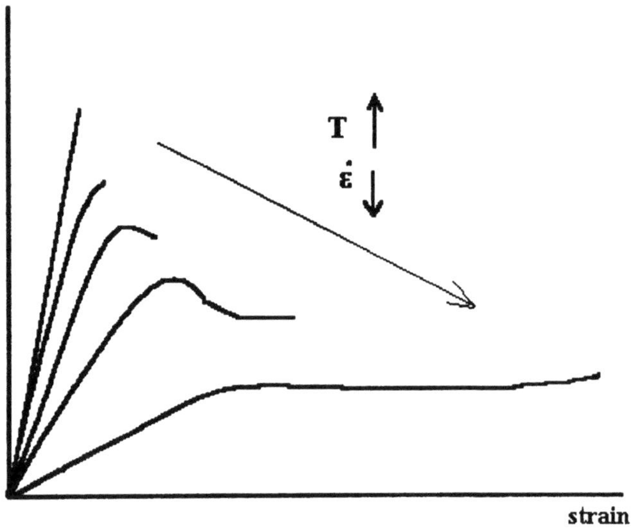

Fig. 10.

uid within a temperature interval of 100–200 °C (Fig. 10). In order to position polymers within the framework of classical mechanics between elastic solids and viscous liquids, the term viscoelastic materials is used. It seems reasonable to establish a clear and satisfactory phenomenological, macroscopic description of the mechanical response of thermoplastics first, before attempting any microscopic interpretation.

The chemical structures of long-chain molecules, their molecular weight and molecular weight distributions, backbone chain stiffness, shape of the molecule, molecular orientation and intermolecular interactions are the most important microscopic parameters affecting the mechanical response of glassy thermoplastics at low strains (elasticity, viscoelasticity). In addition to these variables, at large strains, defect distribution and presence of morphological imperfections control the onset of plastic deformation (yielding, crazing) and fracture [92]. Unlike in glassy thermoplastics, morphology is the single most important structural variable controlling the mechanical response of semicrystalline polymers [93]. Temperature, strain rate, loading geometry, state of stress within the solid, environment, annealing and physical aging are the external variables affecting the mechanical response of thermoplastics.

5.1 Deformation Response of Semicrystalline Thermoplastics

The importance of spherulites and spherulitic morphology for the mechanical response of semicrystalline thermoplastics rests on experimental evidence that

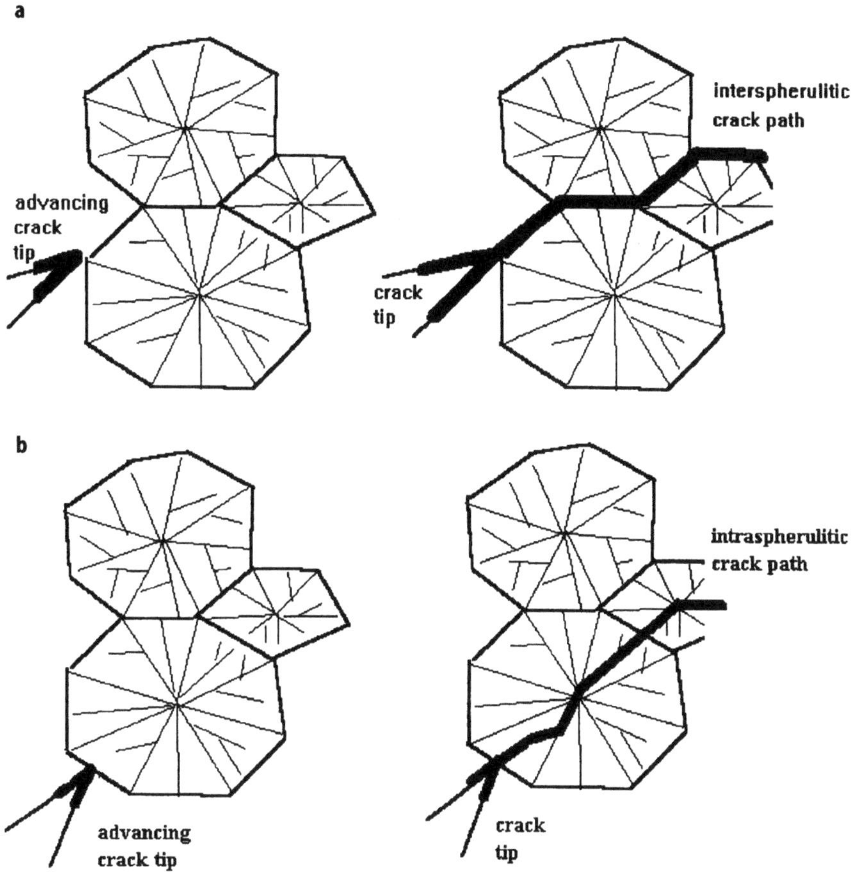

Fig. 11.

spherulite sizes fix the scale for coordinated local mobility of an assembly of segments or, in other words, for the extent of plastic deformation [94]. The deformation response of semicrystalline polymers depends also on the morphology of the crystalline phase, especially on the size and shape of the spherulites. Polymers with small, uniform spherulites prepared by quenching polymer melts exhibit ductile behavior, while morphologies with coarse spherulites resulting from slow crystallization fail in a rather brittle manner [95]. In the latter case, cracks propagate generally along the boundaries of large spherulites. When cracks meet spherulites head-on at boundary intersections, they tend to propagate through these spherulites along surfaces of locally radial lamellae (Fig. 11). This behavior is, however, dependent on the ease of interlamellar cleavage. Polymers with fine spherulitic morphology generally exhibit plastic yielding, though not always in the same manner. In the case of a lower degree of crystallinity, these polymers undergo affine deformation whereas specimens with a high degree of crystallinity generally yield in a heterogeneous manner forming a neck [96]. Observed elongations within the neck of up to 500% or more are facilitated by the ease of chain extension from the neatly folded conformations, and the localized character of the deformation is largely attributable to the manner in which chain-folded crystals respond to stress.

According to current studies, local heating and consequent melting of crystallites seems to play a part only at high strain rates, especially during impact loading. The interpretation of the differences between deformation response of fine and coarse spherulitic structures is based on the fact that it is the spatial arrangement of crystalline and amorphous domains as well as the density of extra- and intraspherulitic tie molecules bridging the amorphous regions and allowing for the load transfer between crystallites, which controls the mechanical response of these complex, heterogeneous solids. At very small strains below 1% and at $T>T_g$ of the amorphous fraction, deformation inside the solid semicrystalline polymer is confined to the disordered amorphous regions [97,98] due to their low stiffness above their T_g. As a result, the elastic modulus of semicrystalline polymers increases with the increasing degree of crystallinity and with the increasing spherulite size [99,100] (Fig. 12). Inside the sherulites, rotation of la-

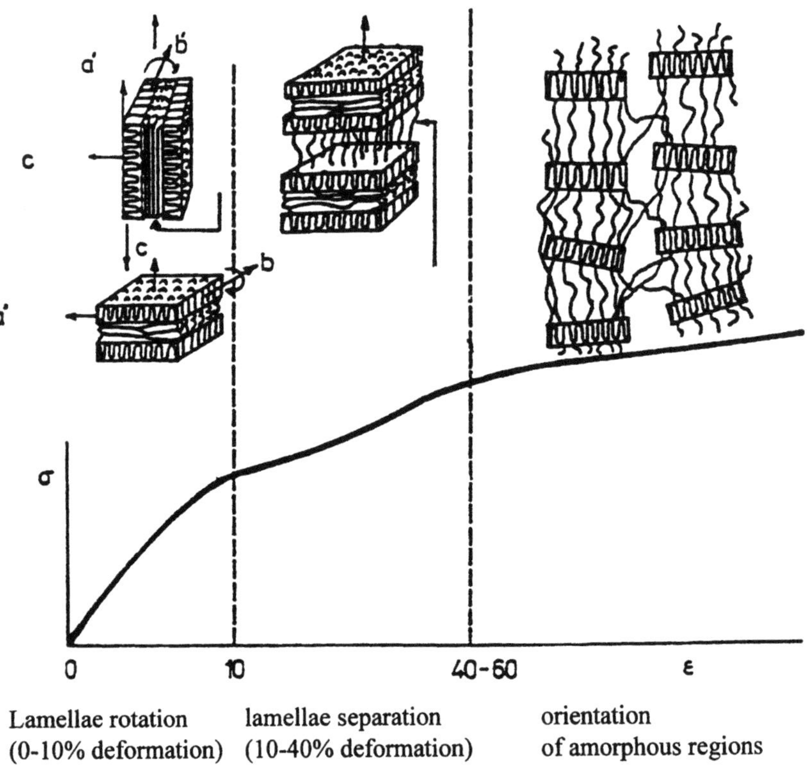

Fig. 12.

mellae occurs within this deformation region [94,101]. The resulting orientation of lamellae depends on their position within the spherulite in respect to the orientation of the external deformation (equator, pole) [102–104].

At larger strains, after disintegration of spherulites, crystalline lamellae undergo plastic deformation processes manifested by macroscopic yielding. Lamellae under tension in their own planes fracture almost immediately with the microcracks being bridged by fine microfibrils 10–30 μm in diameter formed by extended chains pulled out of their folded conformations [65]. As deformation continues, other molecules become extended and feed into these microfibrils, large elongations being achieved before original crystals are consumed [106]. As a result, morphology comprising of stacks of crystalline lamellae oriented obliquely to the tensile stress axis is formed. The total fibrilation can eventually consume the whole crystal when strained at extremely low strain rates and molecules are sufficiently long or there are stronger intermolecular attractions than the weak van der Waals forces [84,105,106]. A small size spherulitic structure has more evenly distributed stresses resulting in more even loading of individual chains. At very large deformations with draw ratios greater than 5, a highly oriented fibrilar structure is formed consisting of microfibrils clumped together into coarser fibrils bridged by unoriented microfibrils crossing longitudinal voids between them [106,107].

6
Elasticity of Thermoplastics

6.1
Phenomenology of Elasticity

In general, Hooke's law is the basic constitutive equation giving the relationship between stress and strain. Generalized Hooke's law is often expressed in the following form [20, 108]:

$$\sigma_{ij} = C_{ijkl}\,\varepsilon_{kl} \tag{2}$$

where σ_{ij}, C_{ijkl} and ε_{kl} are the tensor of stress, the tensor of elastic constant and the tensor of deformation, respectively. Equation 2 has been derived for materials with energy driven behavior, such as metals. The elasticity of polymers as entropy driven materials deviates somewhat from the behavior of ideally elastic solids in several ways. Firstly, deformation of a polymer is strongly dependent on its history and test conditions such as strain rate, temperature, pressure, etc.. As a result, time or frequency should be added as a variable into the constitutive equation (Eq. 2). Secondly, unlike in metals, not all the deformation within the elastic region of polymer behavior can be recovered (duration of the test becomes a test variable). Thirdly, unlike in ideal elasticity, the constitutive elasticity equations for polymers are generally non-linear. Fourthly, the stress and strain used in Hooke's law are defined for polymers in their original sense only at very small deformations. For larger deformations, these parameters must be defined in a more general fashion. And last but not least, polymers can become anisotropic or even orthotropic upon drawing, which greatly complicates the mathematical form of the generalized Hooke's law which has to be emloyed.

6.2 Polymer Elasticity on the Molecular Level

Attempts to establish transformation algorithms between phenomenological and microscopic theories have met with various complications, the principal being the need to distinguish between the response of glassy and semicrystalline polymers principally. The relationships between constitution, configuration and attainable conformations of individual macromolecules and the elastic behavior of a solid glassy polymer vary substantially with the change in temperature. In general, three distinct regions can be recognized, i.e. a glassy state, a glass transition region and a rubbery region. As a result of the long-range order existing in semicrystalline polymers, elastic behavior of semicrystalline thermoplastics is far less dependent on temperature, compared to glassy thermoplastics. Unlike for glassy thermoplastics and thermosets, to date there is no proof of any direct correlation between molecular parameters and elastic properties of semicrystalline polymers. Martin et al. [93] clearly demonstrated that the elastic response of semicrystalline thermoplastics is controlled directly by their morphology rather than by their molecular structure. The effects of chain constitution, configuration, molecular weight and molecular weight distribution on the elasticity of semicrystalline thermoplastics are always indirect and through their effects on the polymer morphology.

The molecular theory of polymer viscoelasticity rests on the work of Bueche [109], Rouse [110], and Zimm [111] who have investigated the behavior of dilute solutions of linear polymers. The molecular theory of viscoelasticity has not been very successful in describing the viscoelasticity of solid polymers over the whole temperature interval and, thus, a modified theory of rubber elasticity has to be employed to describe the elastic response of amorphous polymers above their T_g. On the other hand, it seems reasonable to expect that well below the T_g, the entropic character of polymer elasticity yields to the more energy driven elasticity of polymer glasses. Frozen-in conformations below T_g allow the assumption of an energy driven elasticity with the values of the Young modulus of glassy polymers lower than those for an ideal polycrystalline polymer dicsussed below. Various linear glassy polymers such as PC, PMMA, PVC, PS exhibit Young's moduli within the interval 2–4 Gpa suggesting that the disordered supermolecular structure with no translational symmetry of potential fields resists the external forces only to the limit given by intermolecular interactions. The expression for the Young modulus of elasticity for a system on *N* macromolecules each consisting of *z* equivalent segments can be expressed as follows [112]:

$$E_r(t) = 3NKT\sum_{p=1}^{z}\exp\left(-\frac{t}{\tau_{max}}p^2\right), \tag{3}$$

where *NKT* is the elastic constant of the molecular "spring" at temperature *T*, Poisson's ratio is assumed to be 0.5 and τ_{max} is the maximum relaxation time of the molecular dash pots. The pure elastic modulus is equal to the E_r at t=0:

$$E_1 = 3zNKT \tag{4}$$

where *zN* is the amount of equivalent segments in cubic centimeter. Qualitatively, Eq. (3) predicts an increase of the elastic moduli with increasing polymer density, segment stiffness and size. Additionally, in general, strong intermolecular interactions result in higher elastic moduli.

The molecular interpretation of elasticity of semicrystalline polymers is complicated even more due to their two-phase morphology and the complex manner in which amorphous and crystalline domains are interconnected. Generally, the behavior of the amorphous fraction resembles the behavior of glassy polymers and the elasticity of the crystalline regions is of the same nature as the elasticity of "atomic" crystals. Internal segment rotations are restricted in the polymer crystal and the internal segment mobility consists of simple vibrations about the equilibrium position resulting in energy driven ideal elasticity. Moreover, the elastic properties of crystals are strongly anisotropic. For the PE monocrystal, the Young modulus in the direction of polymer chains is about 300 GPa while the Young modulus measured in the perpendicular direction reaches only about 3 GPa [113]. Similar values for the longitudinal and transverse elastic moduli have been found for other semicrystalline thermoplastics such as PP and PA. The random spatial orientation on crystalline domains within a semicrystalline polymer solid is the cause of the elastic moduli of common semicrystalline plastics being even lower than the transverse moduli of the monocrystals. Preferential orientation of the crystalline regions and orientation of tie molecules within the amorphous domains in highly drawn polymer fibers is the reason for these fibers to reach elastic moduli of the order of 200 GPa, which is about 70% of the longitudinal moduli values for a monocrystal.

In order to interpret the large difference between the longitudinal and transverse moduli, one has to keep in mind that in the direction of chains, covalent bond angles and their lengths would have to be changed in order to extend the long-chain length requiring large forces. On the other hand, substantially lower force is neccessary to move the chains relative to each other, i.e. against the weak van der Waals intermolecular interactions. Since the long-chain molecules in polyamides are interconnected with stronger hydrogen bonds, transverse modulus of PA6 is about three times greater than that of PE, when measured in dry conditions. On the other hand, the planar zigzag conformation of the extended PE chain contributes to its higher longitudinal elastic modulus compared to helix shaped PP chains.

In the PP monocrystal, the helix configuration of the long-chain molecules built in the lamellae topology results in a lowering of the longitudinal elastic moduli compared to the planar zigzag configuration of PE molecules, since the need for high forces to change the covalent bond angles and lengths is reduced to the much lower force demanding a rotational twisting/rocking movement of the segments. As a result, PP crystals exhibit a longitudinal elastic modulus of only about 40 GPa. The transverse elastic modulus is not affected by the helix configuration of the chain, since the intermolecular forces in PP are of the same strength as in the PE and, thus, PP has a transverse modulus of about 2–3 GPa. In semicrystalline polymers exhibiting stronger intermolecular interactions, such as polyvinylalcohol and polyamides, transverse elastic moduli of up to 9 GPa have been observed [113].

In reality, the morphology of a polycrystalline thermoplastic consists of spherulites which holds for common PP, PE, PA 6, PA 6,6 and PEEK crystallized under common conditions. Some semicrystalline polymers as well as the above mentioned moderately filled ones may exhibit lamellar crystalline morphology without any spherulitic order. As a result of random orientation of individual crystallites in spherulites and the manner of their connectivity, the elastic modulus of about 10 GPa has been extrapolated for a hypothetical ideal polycrystalline PE containing no amorphous phase from the dependence of the elastic modulus of PE on the degree of crystallinity. The presence of an amorphous phase which reduces the content of the crystalline phase results in a further reduction of the overall elastic modulus of the semicrystalline polymers compared to ideal monocrystals.

7 Yielding of Thermoplastics

7.1 Phenomenology of Yielding

In a similar fashion to metals, a large number of thermoplastics can develop macroscopic plastic deformation as a response to mechanical loading. The onset of macroscopic plastic deformation is commonly termed the yield point. The viscoelastic-plastic nature of the response of polymers to mechanical loads makes identification of the onset of true plastic deformation very difficult and very often rather arbitrary definitions are used. One has to keep in mind that polymers in general show necking, cold drawing, brittle fracture and large homogeneous deformations depending on the exact loading conditions (strain rate, sample size and shape, mode of loading, temperature, pressure, etc.). Due to the phenomenological nature of the yield point, this fact is a general one, with no respect to their chemical and physical structure [20].

The phenomenological analysis of the stress state at yielding results in a conclusion that the shear yielding is caused by deviatoric components of the stress tensor. This distinguishes shear yielding from crazing caused by a triaxial stress state [26]. The shear yielding deformation occurs without significant change in volume, by changing the shape of the deformed volume element. The *intrinsic* yield point is independent of the test geometry used and it is always followed by strain softening [116–122]. All isotropic glassy polymers exhibit strain softening. Some of them, however, orientate very rapidly and the resulting strain hardening can obscure this process even under compressive or pure shear loading conditions. Similarly, some semicrystalline polymers do not exhibit strain softening due to the onset of orientational strain hardening masking any possible signs of strain softening under usual test conditions [105,123–125]. In standardized tests, σ–ε plots are most commonly obtained from the testing machines with the first maximum corresponding to an instability in deformation and neck formation. This point if often called the *extrinsic* yield point and it depends on the loading geometry. Quite often, extrinsic and intrinsic yield points are very close to each other and the extrinsic yield point is referred to as the yield point in the majority of literature on the deformation behavior of polymers.

Since the test geometry plays a substantial role in the determination of the yield point, it appears reasonable to comment on some of the most common tests. The geometrical features of tensile tests promote extrinsic instabilities causing necking and, thus, they are less favorable for investigations of the true yielding behavior of polymers compared to more favorable compressive loading. Compressive loading eliminates the presence of triaxial tension sufficient to cause craze development and compression also promotes the occurence of an intrinsic yield point. If tested under uniaxial tensile loading, necking and cold drawing may result which are accompanied by a non-uniform distribution of stress and strain along the length of the test specimen. In this case, most deformation is concentrated within the neck.

It is always very useful to be able to predict at what level of external stress and in which directions the macroscopic yielding will occur under different loading geometry. Mathematically, the aim is to find functions of all stress components which reach their critical values equal to some material properties for all different test geometries. This is mathematically equivalent to derivation of some plastic instability conditions commonly termed as the yield criterion. Historically, the yield criteria derived for metals were applied to polymers and, later, these criteria have been modified as the knowledge of the differences in deformation behavior of polymers compared to metals has been acquired [20,25,114,115].

The simplest yield criterion is that of Tresca. This criterion states that yielding will occur when the maximum shear stress on any plane in the tested solid reaches its critical value [20]:

$$\left|\sigma_1 - \sigma_3\right| = 2\tau_2 = \sigma_y \tag{5}$$

where σ_1, σ_2 are principal stresses, τ_y and σ_y are the yield stress in pure shear and the yield stress in pure, uniaxial tension, respectively. More accurate and most widely used is the von Mises yield criterion since the third principal stress component, σ_3, can also be taken into consideration [20]:

$$(\sigma_1-\sigma_2)^2+(\sigma_2-\sigma_3)^2+(\sigma_3-\sigma_1)^2=6\tau_y{}^2 \tag{6}$$

Lower values of the yield stress measured in tension compared to those measured in compression suggest that the effect of pressure, which is important for polymers, is not accounted for in this criterion. Hence, appropriate correction has to be made in order to account for the effect from external pressure. The most frequent version of pressure-dependent yield criterion is the modified von Mises criterion [20]:

$$(\sigma_1-\sigma_2)^2+(\sigma_2-\sigma_3)^2+(\sigma_3-\sigma_1)^2=6(\tau_y-\mu p)^2 \tag{7}$$

where μ is the friction coefficient of the polymer and p is the hydrostatic pressure.

In the case of compressive loading, the Coulomb yield criterion is often utilized in the form [20]:

$$\tau=\tau_c+\sigma_N\tan\phi \tag{8}$$

where τ is the shear stress, σ_N is the normal stress and the angle ϕ is determined as:

$$\Theta=\frac{\pi}{4}+\frac{\phi}{2} \tag{9}$$

where θ is the angle between the normal to the yield plane and the direction of applied external loading.

The measured yield strength of polymers depends substantially upon test temperature and strain rate, expressing the underlying viscoelastic nature of solids made of long-chain molecules. Similarly to the elastic modulus, the yield strength, σ_y, increases with decreasing temperature and increasing strain rate, ε. The Arrhenius type of analysis, expressed in the Eyring model of flow, is usually used to describe this dependence [126–128]. Based on the Eyring theory, Ishai and Cohen [129] expressed the temperature and strain rate dependence of the yield strength of polymers sufficiently below their softening point (T_g, T_m) in the form:

$$\sigma_y=\frac{E^*}{\gamma}+\frac{RT}{\gamma}\ln\frac{\dot{\varepsilon}}{A} \tag{10}$$

where A is a constant for a given polymer, R is the universal gas constant and γ is the activation volume. Despite very good linearity in the σ_y/T vs. $\log\dot{\varepsilon}$ plot, the proposed interpretation of the true physical meaning of the activation energy, E*, and the activation volume, γ, has still not been unambigiously accepted [115]. Experimental and theoretical evidence suggest that plastic flow is a cooperative motion of a large number of equivalent chain segments. This might be the reason why E* does not correspond to any molecular relaxation energy and γ is substantially greater than the size of the statistical random link by 2 to 20 times.

7.2 Molecular Interpretation of the Homogeneous Yielding

Due to the principal difference between the ways in which the cooperative motion of large assemblies of segments or, in other words plastic flow, can be realized in glassy and semicrystalline polymers, molecular theories of yielding are divided in respect to polymer morphology [26]. In addition, plastic deformation of a solid can be divided into homogeneous yielding, in which the whole volume is subject to the same level of displacement, and heterogeneous yielding, in which the deformation is contained in a small, localized volume of the solid. The former is easier to describe mathematically compared to the latter. On the other hand, except in a very few reported cases, the latter is the only one encountered in practical experiments due to the presence of defects concentrating loads and causing development of localized, inhomogeneous plastic deformation.

7.2.1 *Glassy Thermoplastics*

The original yielding theory suggested that the glass transition temperature T_g is reduced upon action of an external stress. At the onset of yielding, the local T_g is assumed to become equal to the test temperature allowing for free relative translational movement of segments within the affected volume [130]. The measured temperature increase within the plastically deforming region apparently supported such a hypothesis. However, contrary to the expected increase in free volume upon yielding, experiments revealed a small decrease in volume at the onset of yielding and confirmed that a further flow after the onset of yielding occurs at practically constant volume [131] proving this hypothesis invalid.

More successful attempts to interpret yielding on a molecular level were based on an extension of the Eyring phenomenological flow theory by incorporating molecular characteristics [132,133]. The modification is based on changes in distribution of rotational conformation states of segments upon stress action and the effect of temperature on them.

Bowden et al. [117,134] and Argon and Bessonov [135,136] have suggested alternative flow theories based on the thermally activated motion of segments. The former is based on an analogy with the theory of plastic deformation in metals using movement of hypothetical dislocations as a basic element of plastic deformation. The strain field of the sheared regions is assumed to be the same as those of dislocation loops with Burgers vector equal to the shear displacement. The original concept of dislocations developed for metals requires translational symmetry in the solid, i.e. the crystal structure, to exist in order to define the term dislocation [123]. Thus, it has to be stressed that the term dislocation in Bowden's theory is only an analogy to that used to describe plastic deformation in crystalline metals. The Argon and Bessonov theory assumes that yielding occurs by thermally activated production of local molecular kinks. This is mathematically treated as a formation of wedge disinclinations. No a priori assumption has to be made in regard to the presence of dislocations in this case. These theories can also be applied to the yielding of glassy thermosets [137].

7.2.2
Semicrystalline Polymers

A two-phase model is often considered for the analysis of the mechanical response of semicrystalline polymers. This model simply assumes the presence of crystalline domains of different shape and orientation in an amorphous glassy matrix [138,139]. One can imagine that for a relatively low degree of crystallinity such a model is close to reality; however, this concept is far from reality for the majority of commercially important semicrystalline polymers exhibiting a degree of crystallinity greater than 40%. Above its T_g, the amorphous "matrix" is assumed to have a rubber-like character, below the T_g it behaves as a glass. The crystalline regions are supposed to undergo shear flow by slip, twinning or martensitic transformation [140]. The deformation of crystalline domains is directionally dependent and rotations of crystallites before their yielding are assumed to take place. For PE, Young [141] has studied the temperature dependence of the slip in directions parallel to the chain direction. Unlike in Bowden's model of homogeneous yielding of glassy polymers, true dislocations exist in the PE crystals [142]. It was found that the plastic deformation of thin PE crystals can be initiated by thermal activation of screw dislocations with the Burgers vectors parallel to the chain direction in a chain-folded crystallite.

7.3
Heterogeneous Yielding

In practice, homogeneous plastic deformation of polymers occurs very rarely. Most frequently, surface defects, local changes in the specimen's cross section, etc., cause localization of strain leading to a more rapid increase in the local plastic deformation compared to the rest of the solid. There are principally two reasons for such a plastic instability to occur. The first reason is a geometrical inhomogeneity in the form of variation in the cross section or shape of the solid resulting in necking when subject to uniaxial tension [143]. The second reason consists of a strain softening taking place non-uniformly in a localized volume of the solid. If strain softening is followed by strain hardening, as commonly observed for semicrystalline thermoplastics, stabilization of the localized plastic deformation can take place also resulting in necking.

As mentioned above, necking is a typical macroscopic manifestation of a geometrical inhomogeneity. This instability occurs because of the stress being concentrated at some point of the solid. The yield stress is, therefore, attained at this point earlier than in the rest of the solid. Resulting localized plastic flow causes orientation of molecules in the neck region resulting in a strain hardening. This allows for a stabilization of the neck and its extension through the whole solid. Increase in test temperature helps the neck development. A more fundamental reason for inhomogeneous plastic deformation to occur is strain softening which may occur after the yield point. Strain softening is an intrinsic property of polymers. It can be described as a substantial reduction of a resistance to plastic deformation after reaching the yield point. If this occurs locally, the material in this region will undergo an even larger deformation than the rest of the specimen. In some cases, orientation of molecules followed by strain hardening can stabilize this process and lead to a neck development and extension. Types of inhomogeneous deformation are discussed in more detail in the book by Kinloch and Young [26].

7.3.1
Glassy Polymers

Crazing and shear banding are the most frequent forms of inhomogeneous plastic deformation in glassy polymers. Shear banding develops in glassy polymers in the case when crazing is suppressed. This occurs mostly in the case of compressive loading [144–146]. As mentioned previously, strain softening is an intrinsic property of glassy polymers and its localization due to molecular heterogeneities is almost inevitable, resulting in heterogeneous plastic deformation.

Crazes develop mostly in glassy polymers, even though some reports have been published attempting to prove crazes in semicrystalline thermoplastics [147]. Crazes are formed by a coalescence of microvoids formed at points of high dilatational stress concentrations such as various structural defects, flaws, surface scratches, sudden changes in cross section, molecular heterogeneities, etc. Low temperature, high strain rates, aggressive environments and plane strain conditions promote crazing [26,148–150].

Craze initiation is related to the generation of microvoids caused by localized plastic deformations occuring near stress raisers. The mechanism responsible for localized plastic flow is shear microbanding. The dilatational character of crazing led to initiation criteria based on either dilatational stresses or critical strains. The most generally accepted molecular crazing criteria are those proposed by Argon [151,152], Haward [153] Andrews [114] and Gent [154]. Argon has suggested that a stable, thermally activated cavitation at loci of large stress concentrations produces microvoids about 10 nm in diameter. Plastic expansion of these microvoids creates craze nuclei when a critical local porosity is reached. This critical porosity requires the presence of sufficient dilatational stresses to produce void interaction. The model in its mathematical form agrees reasonably well with a large volume of experimental data. Haward [153] and Andrews [114] have suggested alternative approaches to craze initiation based on the assumption that the local stress must be sufficient to provide the surface free energy for void nucleation.

Since the molecular crazing criteria require a substantial amount of detailed information about the molecular structure of the solid polymer and no clear correlation to the macroscopic phenomena observed experimentally exists, phenomenological criteria analogous to those for shear yielding were proposed. The most frequently referenced criteria are those of Sternstein and Ongchin [155] and Bowden and Oxborough [156].

The craze growth is a much better understood phenomenon compared to craze initiation, and the theory of Argon and Salama [157] is generally accepted for a description of this process. Craze growth proceeds by the meniscus instability process. Strong experimental evidence supporting this model has been provided by Donald and Kramer [158]. The mechanism of craze growth by meniscus instability results in a growth rate dependent only on the maximum principal stress. It appears that the craze growth is a molecular relaxation controlled process in both air and liquid environments [159]. In a liquid environment, when the craze growth rate exceeds that of the liquid transport to the craze tip, crazing becomes a fluid-flow controlled process.

A gross localization of plastic deformation in craze fibrils, bridging the craze surfaces, makes crazes precursors of brittle fracture. Fracture of craze fibrils reduces the ability of a craze to support load and transforms craze into a crack. Crazes breaking down in a stable manner, i.e. slowly growing cracks, exhibit a stress intensity factor, K_{Ic}, proportional to the n-th power of crack speed da/dt. Since n is proportional to the loss factor, tanδ, the presence of large molecular relaxations helps to stabilize slowly growing cracks. Marshall et al. [160] have derived an expression relating K_{Ic} to the crack speed using the Dugdale zone model.

The molecular interpretation of craze breakdown is based on a concept of fibril rupture due to either chain scission or by disentanglement and plastic flow [149,150]. Despite the involvement of chain scission in the fibril rupture, viscous flow and molecular disentanglement appear to be the principal mechanisms of craze breakdown. It has been suggested [161] that if a majority of the macromolecules passing through the craze/bulk interface terminate in the fibril, separation of the craze surfaces will involve disentanglement of segments rather than rupture of primary bonds. Ball-like objects, representing contracted broken fibrils, observed on the fracture surfaces of crazes are consistent with this explanation. Hence, an increase in molecular weight causes an increase in the ultimate fibril deformation resulting in a greater crack opening displacement, δ_{tc}, of the craze breakdown originated cracks. However, the increase in δ_{tc} also depends profoundly on the chemical structure of the long-chain molecules [162–164]. The molecular structure of the particular glassy thermoplastic affects crazing very substantially when macromolecules are oriented unidirectionally, such as in the craze fibrils extended close to their rupture [165,166]. The density of molecular entanglements and the contour length of a chain between entanglements have an important bearing as to whether the polymer will craze or undergo shear yielding. Additionally, the character of entanglements controls to a great extent the mechanical response of the craze matter, i.e. fibrils [167,168].

Shear yielding in the form of a quasi-homogeneous, bulk process can contribute substantially to the crack resistance of a polymeric solid. On the other hand, however, localized shear yielding in the form of shear micro-bands is believed to be a precursor of brittle fracture in many semicrystalline and glassy thermoplastics. The nature of developed shear bands depends on test conditions and the type of polymer. In PS, a sharp, fine texture of shear bands exists leading to a term "microshear bands". Microshear bands are usually 0.5 to 1 μm thick and they are oriented at an angle of 38° to the direction of the compressive load. On the other hand, shear bands observed in PMMA are wide with poorly defined boundaries resembling broad, diffuse zones, hence termed "diffuse shear bands". More importantly, however, the strain in microshear bands reaches several hundred percent while that in the diffuse shear bands is only a few percent above the strain level in the rest of the material [26]. Some indications suggest that the diffuse shear bands are viscoelastic rather than plastic [169]. Other glassy thermoplastics form shear bands intermediates between microshear and diffuse shear bands. It was also found that the diffuse shear bands are formed by very fine deformation bands of the width of 50 nm. Unlike the 500 to 1000 nm microshear bands, these fine deformation elements of diffuse shear bands are not continuous [170]. On the other hand, the shear deformation in these sub-microshear elements was of the order of the deformation found in microshear bands.

Strain rate, test temperature and the thermal history of the specimen all affect the appearance of shear bands in a particular glassy polymer [119]. The differences in morphology of shear bands was proposed to be due to different rates of strain softening and the rate sensitivity of the yield stress. Microshear bands tend to develop in polymers with a small deformation rate sensitivity of σ_y and when relatively large inhomogeneities exist in the specimen before loading. This is sometimes characterized by a factor e_{sb} introduced by Bowden in the form [119]:

$$e_{sb} = \frac{\left.\dfrac{d\sigma}{d(\ln\dot{\varepsilon})}\right|_{\varepsilon}}{\left.\dfrac{d\sigma_y}{d\varepsilon}\right|_{\varepsilon}} \quad . \tag{11}$$

The parameter has a value of 0.11 for PMMA at room temperature and 0.2 for PS at 80 °C. In both cases, diffuse shear bands develop. Reducing the test temperature to about 70 °C, e_{sb} for PS decreased to about 0.016 and sharp, well defined shear bands were developed.

Interestingly, the contribution of diffuse shear bands to the total deformation of the specimen is large, despite relatively low deformation existing in them. On the other hand, large plastic deformation in microshear bands does not contribute so substantially to the total deformation of the specimen. It appears that a small deformation over a large volume has a much larger effect on crack resistance than a large deformation within a small volume.

Regarding the effects of shear band structure on the fracture mode in glassy polymers, Wu and Li [170] concluded that when microshear bands propagate in a specimen cross section, a shear failure is produced with a very small overall deformation, resulting in a brittle character of the fracture process. This mode of fracture is favored at high strain rates, low temperatures and slow cooling of the specimen during solidification from the melt. Propagation of the diffuse shear bands through the specimen cross section resulted in large overall deformations producing a large kink and the specimen failed in a ductile manner. Studies of the kinetics of shear band propagation resulted in measured activation energies in the order of 270 kJ/mole which is comparable with 251 kJ/mole found for a bulk yielding. Thus, it seems that the propagation of shear bands is controlled by plastic strain rate at the band tip.

Localization of shear deformation into narrow bands is now believed to be the precursor of brittle failure in thermoplastics and in filled polymers [26]. The experimental evidence for this hypothesis lays in observations of microvoid formation at the intersections of shear bands [171,172]. It has been proposed that stress raisers such as air bubbles, flaws, defects, molecular inhomogeneities, filler particles, etc., enable microshear banding to occur at external loads well below the bulk yield stress is reached. When two microshear bands intersect, microfibrils from each of these bands are subject to further straining. In this extremely strained area, chain disentanglement or, eventually, even chain scission takes place preferentially compared to the rest of the shear bands. This process will result in a generation of microvoids which can act as nuclei for crack growth initiation. An additional mode of failure resulting from the presence of sharp shear microbands is the rupture of microfibrils in these bands [173–175]. Kinloch and Young [26] have summarized the observed sequence of failure processes leading to a brittle fracture of glassy polymers in the following steps: (i) strain softening in the plane strain conditions causes microshear bands to develop, (ii) growing shear bands intersect, (iii) microvoids form at the intersection of shear bands nucleating a craze structure and, (iv) craze breakdown and crack initiation occur. The presence of sharp notches, thick specimens, low temperature, thermal annealing and high strain rates will promote this type of fracture.

Crazing and localized shear yielding are not mutually exclusive deformation processes. Both these mechanisms were observed simultaneously in many polymers [176,177]. It is natural to expect that an interaction between crazes and shear bands will occur as they grow through the specimen cross section. Unlike some early theories, current models on interactions between crazes and shear bands suggest that the only possibility for a shear band to terminate craze growth is in the case when the shear band is formed at the craze tip [177–185]. The originally proposed explanation based on an assumption that an interaction between a craze and an existing shear band can lead to a stress relief at the craze tip was proven invalid because of the large strain already existing in the shear bands. The more probable result of this interaction would be a premature craze breakdown and brittle crack nucleation rather than the proposed craze stabilization.

7.3.2
Semicrystalline Polymers

The two-phase morphology of semicrystalline polymers causes a substantially different response to mechanical excitations resulting in an inhomogeneous plastic flow on microscopic and macroscopic levels different from those observed in glassy thermoplastics [123]. Moreover, the deformation of crystalline regions is anisotropic resulting in slips and twinning in only certain crystallographic directions. It is, however, common that no shear bands develop when the morphology of a semicrystalline polymer is satisfactorily isotropic. Interactions of localized shear bands, resulting in formation of microcracks, is also believed to be a precursor of brittle fracture in semicrystalline polymers. On the other hand, large kink bands are observed in deformed oriented semicrystalline polymer fibers [186]. It is thought that these kinks are formed by combination of a slip in the crystalline regions and a plastic flow in the amorphous regions. The angle of the kinks in respect to chain orientation depends on the chemical structure of the long-chain molecules and the conditions of the test.

Semicrystalline polymers deform elastically by affine deformation up to the relative total deformation of about one percent. The two-phase morphology and relatively low shear strength of lamellae results in an onset of plastic deformation at very low strains. In a macroscopically quasi-isotropic polymer, orthotropic lamellae are randomly oriented in respect to the external loading direction between the two extreme orientations, i.e. parallel and perpendicular, due to their radial growth pattern in a spherulite as well as due to the helix topology. In the course of plastic deformation, rotation and shear deformation of lamellae occurs within the crystallites and bunches of lamellae followed by increasing inclination of chain orientation in respect to the lamela plane. Eventually, lamellae disintegrate into small blocks of folded-chain morphology joined by tie molecules. As a result of lamellae breakage, the concentration of tie molecules increases with increasing deformation. Blocks of folded chains held together by the tie molecules form microfibrils with orthotropic symmetry of properties caused by orientation of chains in crystalline blocks as well as orientation of extended tie molecules in the amorphous phase. The diameter of these fibrils ranges from 10 to 30 nm and their length may be in the order of micrometers. The nature of this deformation process dictates that the tie molecules be concentrated on the surface of individual microfibrils with a portion of them bridging the gap between neighboring microfibrils, ensuring integrity of larger bunches of microfibrils – fibrils. Orientation of molecules in the amorphous phase is one of the most profound changes of morphology occuring in microfibrils compared to an undeformed polymer.

8
Fracture of Thermoplastics

8.1
Phenomenology of Polymer Fracture

Brittle fracture is considered the most dangerous type of failure in any engineering or load-bearing application of polymers. Flaws of various kinds existing in any polymeric solid cause localized stress concentration resulting in local overloading of the material and consequent premature fracture. The stress concentration factor associated with a microcrack of length 2a having a very small radius r at the tip, positioned perpendicularly to the applied stress direction, was expressed by Inglis [92] as $2(a/r)^{1/2}$. It is, thus, of pivotal importance to determine the critical value of the external stress, σ_c, which causes a brittle crack to propagate catastrophically through the specimen cross section. Griffith [272] has derived an expression relating σ_c to materials characteristics and flaw size for brittle glass. This has been modified since by Orowan and Irwin [92] using the critical strain energy release rate, G_c, instead of the surface enrgy, γ, used originally by Griffith, into the form:

$$\sigma_c = \left[\frac{2E(G_c + 2\gamma}{\pi a}\right]^{1/2} \tag{12}$$

where E is the tensile modulus of elasticity. In common thermoplastics, surface free energy is three orders of magnitude lower than and, thus, it can be neglected (Table 9).

Brittle fracture of glassy polymers generally starts by a breakdown of a large craze causing initiation of an unstable crack. In order to obtain more ductile failure in glassy thermoplastics, generation of a large number of small, stable crazes instead of one large craze must be achieved. The preferred way of generating this type of failure is by incorporating controlled dispersion of stress raisers of de-

Table 9. Critical strain energy release rates, G_c, for common thermoplastics measured under impact loading promoting brittle fracture [92]

Polymer	G_c (kJ/m^2)
PS	0.6–0.8
PMMA	1.4
PVC	1.4
HDPE	3.1
PC	4.8
PA 6,6	5.0
Medium density PE (ρ=0.94 g/m^3)	8.4
HIPS	14
LDPE	34
ABS	47

sired size and properties. Elastomer inclusions with diameters of micron size well bonded to the original glassy matrix are the most frequently used intentionally introduced stress concentrators. Commercial viability of this approach has been proven in HIPS, ABS and rubber modified PP.

At small and moderate strain rates, fracture in semicrystalline polymers occurs in the process of cold drawing prior or during deformation strengthening. In the course of cold drawing, the primary macroscopically isotropic properties of a semicrystalline polymer are altered into orthotropic ones due to the transformation of the spherulitic morphology into the microfibrilar one. Microfibril ends form point vacancies or dislocations representing the weakest points of this morphology and are most probably the loci for crack initiation. These defects grow with increasing strain either perpendicularly to the draw direction by a fracture of neighboring microfibrils or parallelly to the draw direction up to the locus of a similar defect in the structure of the neighboring microfibril allowing expansion of the defect in the direction perpendicular to the draw direction. The former mechanism of crack growth is more common in polymers with strong intermolecular interactions (nylons) while the latter is common in polymers with weak intermolecular interactions (polyolefins). Both processes proceed initially in a stable manner resulting in an increased stress concentration in the vicinity of the growing defect. As soon as the growing crack reaches its critical size, catastrophic failure occurs resulting in the failure of the solid. Fracture surfaces resulting from the two mechanisms of crack initiation differ substantially. Smooth, glass-like fracture surfaces are observed for the first mechanism while rough surfaces covered by ends of fractured microfibrils are characteristic for the latter [27].

It has been shown that in many cases, especially at low-temperature fractures, the crack initiation described above contributes to the polymer toughness far less than the deformation processes accompanying its propagation. All three stages of the failure process, i.e. initiation, propagation and termination (important for very thin specimens), dissipate mechanical energy stored in the material during its loading to the point of failure. The distribution of the strain energy dissipated during the individual stages of the fracture depends on both structural variables (glassy, semicrystalline, filled, elastomer modified, molecular weight, etc.) and test conditions (strain rate, temperature, specimen geometry, etc.). The structural variables control the nature and character of the actual dissipative mechanisms on the microscale (shear yielding, crazing, shear banding, etc.) while the test conditions control the relative importance of the respective microdeformation processes (operates the process for which the test conditions are more favorable). Unlike the case of the elastic modulus, fracture is strongly dependent on the state of stress at the crack tip and ahead of it throughout the fracturing solid. Both structural variables, expressed in terms of material properties (E, σ_y, G_c, ν), and test conditions (temperature and strain rate dependence of the material properties, specimen geometry, crack tip radius) are bound together in controlling the state of stress [17,18,92,187,188].

The major contributing dissipative processes at the crack tip are shear yielding in semicrystalline thermoplastics and crazing in glassy thermoplastics.

Shear yielding and crazing are not mutually exclusive processes, and, frequently, both operate at the same time. The relative contributions from shear yielding and crazing will depend on test conditions (temperature, strain rate, crack tip radius. etc.) as well as on the structural variables (crystalline morphology, density of tie molecules, molecular structure, etc.). Frequently, a transition in the major deformation mechanism from shear yielding to crazing or shear banding or vice versa is accompanied by a sudden change in the measured crack resistance. This phenomenon is often termed the ductile-brittle transition (DBT).

The large practical importance of this phenomenon has led to many hypotheses as well as theories attempting to interpret the experimental data and predict effects of both types of variables on the position of DBT and the magnitude of toughness change. Some of the explanations of the DBT have been proposed using the Ludwik–Davidenkov–Orowan hypothesis that the brittle fracture occurs when the yield stress as a material property exceeds a critical value. It assumes that brittle fracture (brittle strength) and yielding (yield strength) are independent processes with different dependences of the respective strengths on the strain rate and temperature (Fig. 13). Generally, yield strength is expected to depend on both these variables in a more profound fashion than the brittle strength. It is argued that whichever process can occur at lower external stress will actually be the operative one. Thus, the intersection of yield and brittle stress defines the position of the DBT where conditions for both types of failure possess the same probability. Since this intersection must depend on T and $\dot{\varepsilon}$, mostly due to the strong dependence of σ_y on these variables, DBT shifts to higher temperatures with increasing strain rate. At very high strain rates, heat formed during yielding is not conducted away due to the poor heat conductivity of polymers causing local isothermic-adiabatic transition preventing any strain hardening and, thus, the ductile fracture prevails again. In agreement with this, an increase in fracture resistance has been observed at moderately high strain rates (lower than ballistic strain rates).

It is generally accepted that the most effective dissipative processes are those involving large volume plastic deformations of the fracturing solid prior to crack initiation. Large plastic deformations also take place in shear bands and large crazes; however, extreme localization of plastic deformations into small volumes of the material leads to macroscopically brittle failure initiated from these areas of large but localized plastic flow. This is even more pronounced during straining of preexisting cracks or notches at high strain rates or at low temperatures. An obvious method of increasing the amount of dissipated energy is an expansion of the volume of a polymer involved in shear yielding or crazing or, in other words, delocalization of plastic deformation. This is most effectively achieved by incorporating a secondary component of prescribed elastic properies, inclusion size and interfacial adhesion to the matrix. Elastomers, thermoplastic inclusions and, in some instances, rigid inorganic particles and reinforcements are the most frequently utilized secondary "toughening agents". The two primary deformation processes encountered in thermoplastics (shear yielding, crazing) are greatly altered by incorporating secondary particulate matter

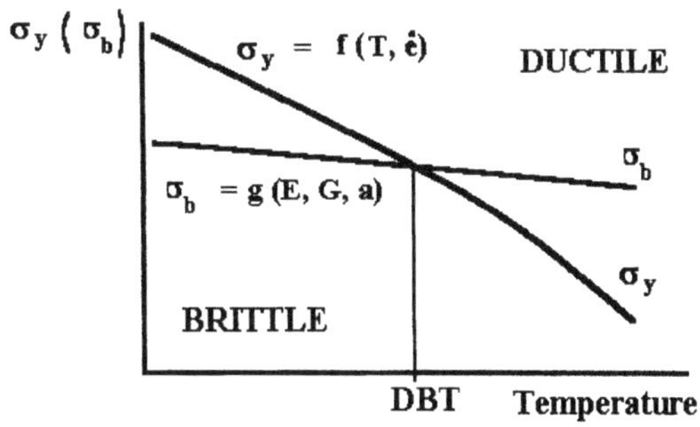

Fig. 13.

into thermoplastic matrices. In addition, completely new deformation/dissipative processes may be introduced due to the presence of particulate inclusions both soft and rigid. The new dissipative processes, as well as redistribution of the contributions from the existing ones, often arises from the presence of the secondary component (filler particles, elastomer inclusions) and from interactions between the host polymer and the secondary particles (interfacial cavitation, particle deformation, etc.). Various aspects of this topic have been reviewed recently in Advances in Chemistry Series 233 [189] and in the books by Karger-Kocsis [190] and Rothon [191].

8.1.2
Delocalization of Shear Banding

Extensive shear yielding at the crack tip is the major dissipative mechanism in many tough polymers. On the other hand, localized shear yielding, such as in shear banding, is believed to be the precursor of brittle fracture in many thermoplastics. It is now well documented that the delocalization of shear banding or, in other words, spreading the shear yielding over a larger volume of the polymer near the crack tip, is a desirable process capable of enhancing crack resistance of these thermoplastics. As put forward above, it can be achieved in many otherwise brittle polymers by incorporationg a uniformly dispersed secondary discontinuous component. This approach has been shown to be especially effective when the elastic modulus of the dispersed inclusions is substantially lower than that of the matrix [5, 26].

The explanation of the effect of secondary inclusions on the delocalization of shear banding is based on the concept of modification of the local stress fields and achieving favorable distribution of stress concentrations in the matrix due to presence of inclusions. This leads to a reduction in the external load needed to initiate plastic deformation over a large volume of the polymer. As a result, plastically deformed matter is formed at the crack tip effectively reducing the crack driving force. Above approximately 20 vol% of the elastomer inclusions, cracks become effectively blunted and the originally brittle failure is transformed into the plastic hinging [188].

The original Goodier solution [192] for the stress field around an isolated particle in a uniformly stressed isotropic matrix resulted in a maximum stress concentration of about 1.9 at the equator of the inclusion [193]. However, it has to be borne in mind that this solution belongs to the class of analytical "single particle solutions", which do not take into account possible stress field interactions or overlaps in multiparticle systems. The solution for interacting particles, i.e. for volume fractions above 0.09, was obtained using the numerical F.E.A. method [185,194,195] showing that due to an overlap of the stress fields of neighboring monodisperse particles in the specific spatial arrangement considered, a stress concentration of up to 6 can be achieved apart from the inclusion surface. The location of the maximum stress concentration moves away from the particle surface with increasing inclusion volume fraction and its actual position depends on the spatial packing of the inclusions. It was also shown that there is not much of a difference in the shape of the stress field around an elastomer inclusion, a void and a rigid inclusion with no adhesion to the matrix [4,188,194,195]. Additional consideration should be given to the fact that a morphology of a semicrystalline matrix can be affected substantially when the inter-inclusion distance becomes of the order of lamella size.

8.2 Fracture of Thermoplastics on Molecular Level

Since an understanding of the importance of any one process contributing to the failure in thermoplastics and the control over these processes is only partly attainable, a knowledge and understanding of the nature of endurance limits is of extreme importance for successful use of plastics, in particular engineered thermoplastics [27]. In terms of the failure type, polymer fracture may occur as a rapid extension of an initial defect, plastic flow of the matter and the thermally activated flow of the macromolecules. In all these cases, however, fracture is a localized phenomenon characterized by a large inhomogeneity of deformations.

The fracture processes on the molecular level are defined in terms of localized physical rearrangements of chains realized by segment rotation, cavitation and slip. In some instances, especially in highly oriented fibrillar polymers, at low temperature and high strain rates, primary bond breakage may be involved in the rupture. This process is commonly termed the chain scission. The relative importance of the particular process in the overall balance of mechanical energy dissipation depends on various parameters such as molecular weight, presence and type of side groups, intra- and intermolecular interactions and it is often affected by morphological factors, such as the crystalline superstructure. In addition, the state of stress, loading geometry, strain rate, temperature, etc. are the most important external variables affecting the nature of the fracture.

The molecular interpretation of macroscopic fracture is obscured due to a large difference in scales between individual long-chain molecules and macroscopic polymer solid subjected to mechanical loads. In addition, it is worthwhile noting that most of the results which provide the background for understanding the fracture on the molecular level were obtained using highly oriented polymeric fibers and, hence, the concepts can be translated into the investigations of brittle fracture of quasi-isotropic polymeric solids only with great caution.

It is now generally accepted that the fracture of thermosets and rubbery materials consists of breaking the primary chemical bonds in their backbone chains. The role of bond breaking appears much less important for thermoplastics. EPR spectroscopy experiments have shown that in quasi-isotropic PE and PP specimens broken at low temperatures, less than 1% of the primary bonds existing on the crack plane were actually broken. Disentanglement of the long-chain molecules was the primary molecular process allowing the crack to propagate. Voids and other crack nuclei can be formed in thermoplastics by thermal motion of chain segments without breaking any primary, load-bearing bonds. Thus, localized chain scission and bond strength determine the fracture resistance of polymers only in the case when no flow occurs. One can expect that loading of individual chains up to the bond strength can take place in thermosets with little or no flow allowed and, in some instances, in highly crystalline, oriented thermoplastics, strained at extremely high strain rates or at very low temperatures.

8.2.1 *Chain Loading*

It was observed experimentally that even in the glassy polymers or in polymer solutions, macromolecules can break by rupturing primary bonds. In order to load the chain up to the strength of primary bonds, intermolecular interactions of the order of those encountered in a crystal and sufficiently large elastic modulus of the chain segments must exist. As intuitively expected, there is a distribution of stresses along the chain with two free ends and a maximum load existing in the middle section of the macromolecule. The axial forces attained in these polymers are, however, limited by the slip of neighbouring chains.

An alternative way of loading chains in a glassy polymer is via the forces of inertia, i.e. by stress wave propagation. For strain rates below 50 s^{-1}, brittle thermoplastics such as PMMA, SAN and PS behave "classically", leading to an increase in strength and a decrease in ultimate strain with increasing $\dot{\varepsilon}$. At strain rates ranging from 50 to 66 s^{-1}, a transition to stress wave initiated fracture was observed. As a result, a tenfold reduction in the load-bearing capacity of the polymer was observed. Similarly, at room temperature, thermoplastic fibers (PA-6, PETP, aromatic polyimide, etc.) behave "classically" in the range of strain rates from 0.1 to 140 s^{-1}. After reducing the test temperature to below −67 °C, a decrease in the strength with increasing strain rate was observed for strain rates above 30 s^{-1}. It has to be borne in mind that the strain rates are considered in the axial direction of the macromolecules [27].

8.2.2
Relief of Axial Stresses in the Chain

Upon loading to a high stress level, chains have to relieve the built-up stresses. The mechanism available to relief the built-up stress include the chain slip, conformation change and chain scission. Generally, in thermoplastics, the relaxation times needed to relax strained chains at room temperature vary from 10^{-3} to 10^{2} s. Thus, the relaxations are faster than the mechanical excitations at slow tests allowing for a relief of a substantial portion of the load for the duration of the test. Rapid loading during an impact test or at the tip of a running crack can bring more non-extended chain segments to high stress levels and, eventually, to chain scission, compared to slow loading, within the same strain interval.

The effect of change in conformation of the chain can be illustrated by the behavior of PE. The transformation of the 4 gauche conformations within a PE segment of a length of 5 nm corresponds to an increase in length by 0.25 nm, i.e. by 5%. This reduces the axial elastic forces in the chain by 0.05×E, which is equal to about 10 Gpa. If the maximum static loading in the PE chain produces an axial stress of 7.5 Gpa, this conformation change fully relieves this load and completely unloads the chain. The rates of conformational changes are relatively high and they tend to increase with the applied external load.

The final mechanism of stress relief is thermomechanically activated chain scission. Primary bond breakage can be homolytic, ionic or by a degrading chemical reaction. It is worthwhile to note that the relative slippage of chains, microfibrils and fibrils reduces or prevents the mechanical scission of chains in quasi-isotropic polymeric solids. In other words, chain scission is an important mode of fracture only in highly oriented thermoplastic fibers or in thermosets.

In PE, a draw ratio of 5 has to be achieved before any measurable bond breakage occurs. The fact that semicrystalline polymers do not exhibit significant chain scission for draw ratios below 3 is in agreement with Peterlin's model of plastic deformation [196]. The inter-spherulitic tie molecules undergo chain scission first, while the intra-spherulitic and inter-crystallite tie molecules are not overloaded in the first steps of loading. They become crucial in the later stages of deformation in the formation of a microfibrilar structure. Chain scission was observed in semicrystalline polymers for draw ratios ranging from 3 to 10. At these draw ratios, the presence of laterally rigid crystalline regions permits static transition of large axial stresses into chains resulting in attainment of critical stress before macroscopic fracture.

With the exception of PC, amorphous, non-oriented polymers did not produce measurable amounts of broken segments when subjected to tension. As has been shown in previous paragraphs, large axial stresses capable of chain scission in amorphous polymers can only be transmitted into the chain by friction of slipping chains requiring strong intermolecular interactions. In addition, macroscopic fracture occurs before a widespread chain overloading and scission occurs, which is opposite to the behavior of semicrystalline polymers.

From numerous experimental observations, it is generally accepted that the morphology of a fracture surface is not simply formed by the rupture of molecular chains across a fracture plane. There is always a plastic deformation at the crack tip, resulting in large plastic flow of the chains within a layer of finite thickness from the fracture surface. Intuitively, one can expect dependence of the extent of plastic deformation on the same external parameters that affect yielding, i.e. temperature and strain rate. Generally, the higher the possibility of intersegmental motion, the larger the extent of plastic flow. Localization of fracture processes in an immediate vicinity of the crack plane emphasizes local, stochastical behavior of chains during the fracture.

8.2.3
Chain Scission and Disentanglement

A hydrocarbon chain is in a constant thermal motion and, without an external force field, the chains fluctuate around the most stable position given by the distribution of possible conformations at the temperature [184]. Gaussian random coil chain configuration describes the average chain end-to-end dimensions satisfactorily. The action of external forces at the ends of a molecule causes displacements of chains from their equilibrium conformations and evokes retractive forces. For a hydrocarbon chain of M_w=14,000, extended length 125.5 nm and the end-to-end distance r=7 nm, the maximum exerted force is 10 MPa. The non-Gaussian treatment provides somewhat different results, however, the level of forces exerted by the random coil macromolecules are much lower than the theroretical strength of the primary bonds. No hindrance to rotations of segments was assumed in these calculations. Generally, the presence of such hindrance reduces the retractive forces by a factor of 1.5 to 2.0. The presence of strong intermolecular interactions, such as hydrogen bonds in polyamides, affects the retractive force substantially, causing a restriction of the number of possible chain conformations. In addition, the transitions between the conformations can be dependent on each other resulting in a cooperative character of these rearrangements. Theoretical calculations performed for extended PE and PA chains at room temperature provided retractive forces small compared to the ultimate forces which can be supported by the chain.

In linear amorphous thermoplastics, large axial forces cannot be applied at a point. In analogy with fibrous composites, the large axial forces in a macromolecule have to be built up gradually along the chain through weak intermolecular forces, i.e. by intermolecular "friction" and shear transmission. The characteristic feature of the loading of a chain in a semicrystalline thermoplastic is the periodicity of the potential field. For the ideal PE lamella, the limiting axial stress was estimated at about 7.5 Gpa. In other words, by being built in the crystalline lamella, the long-chain molecules in semicrystalline thermoplastics can be loaded to a very high level of stress, which is generally two orders of magnitude greater than that achievable in glassy thermoplastics. It has been found that the mechanical excitation decays in the direction of lamella thickness. For an ideal

PE crystal, the penetration length of the mechanical excitation into the lamella was found to be about 5 nm. As a result, if the tie molecule ends in the lamella or continues and ends in an amorphous region, the axial forces it can be exposed to are very small. Such a tie molecule contributes very little to the strength of a semicrystalline solid.

Addition of strong intermolecular interactions such as hydrogen bonds between long-chain molecules, such as those in polyamides, creates a relatively inhomogeneous potential field, because these attractions are stronger at C=O and N–H sites than at CH_2 sites. It was suggested that the hydrogen bonds are responsible for the difference between the cohesion energies for PA-6 (18.4 kcal/mole) and PE (6.3 kcal/mole). These calculations provide the limiting stress to pull a tie molecule from an ideal PA-6 crystal of the order of 22.4 Gpa, i.e. about three times that of PE. Despite relatively crude assumptions and simplifications used in the original model, the results hold even when more sophisticated approaches are applied.

9 Polypropylene

More than 60% of produced polyolefins (PE, PP) have been introduced to the market as compounds, while only about 23% of the volume of other thermoplastics have been used for compounding. Polypropylene is considered one of the primary candidates to become the matrix of choice for engineering new thermoplastic compounds, replacing many small volume engineering plastics, using physical ways to modify its properties via compounding, filling and reinforcing,. Hence, it seems rational to devote special attention to polypropylene as the primary matrix polymer used in filled thermoplastics. The reader interested in greater details going beyond the scope of this volume is referred to the book by Karger-Kocsis [190].

Polypropylene (PP) is a semicrystalline commodity thermoplastic produced by coordination addition polymerization of propylene monomer [197]. Most frequently, stereospecific Ziegler–Natta catalysts are used in industrial processes to produce highly stereospecific crystalline isotactic (iPP) and syndiotactic (sPP) polymer with a small portion of amorphous atactic PP as a side product. Polymerization of non-symmetrical propylene monomer yields three possible sequences; however, the steric effect related to the methyl side group highly favors the head-to-tail sequence. The occurence of head-to-head and tail-to-tail sequences produces defects along the PP chain [198]. Presence of such defects affects the overall degree of crystallinity of PP.

In order to obviate the inherently low fracture toughness of highly crystalline iPP, especially at low temperatures, copolymerization with ethylene is utilized producing a similar distribution of defects along the PP chain. The ethylene content and its position along the chain (random, block) greatly affects the properties of the resulting copolymer such as elastic moduli, elongation to break, thermal stability and the degree of crystallinity [199]. Low ethylene content (up to

5 mol%) in the copolymer results in PP with a reduced degree of crystallinity without substantial reduction of stiffness and substantially enhanced low temperature fracture toughness [198]. Copolymers with higher ethylene content (above 10 mol%) and random distribution of Et along the backbone chain of PP possess the behavior of thermoplastic rubbers (EPR, etc.). This molecular structure prevents any crystallization occuring. On the other hand, if blocks of Et are sufficiently long, ethylene and propylene domains which crystallize separately may result. In order to allow for subsequent cross-linking, non-conjugated diene monomers are used to produce the ethylene-propylene-diene elastomers (EPDM). Recently, two major worldwide suppliers of PP (Dow Chemicals, Exxon Chemicals) have introduced commercial processes based on metallocene catalysts to produce a range of PPs and copolymers of propylene with various other olefins and even, in one process, with polar monomers. These materials are currently used as-polymerized mostly in food packaging and medical disposables applications or as thermoplastic elastomers.

Unlike linear PE having a planar zig-zag chain configuration (Fig. 2), PP forms a symmetrical helix molecule with three monomeric units per single revolution (0.65 nm) [200]. The side methylene groups are alligned in three columns parallel to the helix axis. These -CH_3 columns are shifted relative to each other by an angle of 120° (Fig. 2). Under common conditions, this symmetrical configuration of iPP chains results in relatively uninhibited crystallization in a stable monoclinic α-form [93,201]. At high undercooling or at high pressure, less stable hexagonal β- and triclinic γ-crystal modifications can form [202–205]. Syndiotactic PP which also has a symmetrical molecule crystallizes in orthorhombic form [206]. Typically, commercial iPP contains 40–60 wt% of the crystalline phase with the melting point ranging from 167 to 180°C in relation to the crystal form and lamellae thickness. The equilibrium melting temperatures for the iPP lamellae of the various crystal forms range from 180 to 220 °C for the α-form [207–209] and 176 °C for the β-form [210]. A more detailed description of the current view of the crystalline morphology of PP homo- and copolymers has been published by Cheng et al. [211] and Varga [212] has published a detailed review of the various phenomena related to the crystallinity of iPP. With the relatively new development of the Zr- and Hf-based catalysts allowing production of relatively pure syndiotactic PP on an industrial scale [213], more attention is being paid to investigations of the crystallization of this polymer [206,214–219].

Neat iPP crystallized from the melt exhibits spherulitic morphology of the crystalline phase [220]. In some cases, and under very specific conditions, cylindrites, axialites, quadrites hedrites and dendrites may be formed of iPP [65,220–223]. In general, crystallization from quiescent melts results in spherulitic morphology while crystallization from melts subjected to mechanical loads results in cylindrites [224,225]. Transcrystalline morphology is formed when crystallization takes place on the solid surface of fillers or reinforcements. Movement of fibers in the melt during crystallization produces row nuclei which is the first step in forming cylindrical crystalline structures [226,227]. Crystalline supermolecular structures caused by oriented crystal growth from heterogeneous surfaces is commonly termed transcrystallinity [228]. Transcrystallization takes place when the density of the crystal nuclei is substantially greater on the surface of solid inclusions than in the melt bulk. Since polyhedral spherulites cannot develop due to restricted lateral growth, crystallites are allowed to grow only in stacks perpendicularly to the surface plane [40,229–232].

Spherulites up to 0.2 mm in diameter are built in iPP from crystallites formed of twisted lamellae typically 10 nm thick and 1–10 μm wide [233–235]. Theoretical descriptions of spherulite formation can be found in the literature [236,237]. Spherulites of PP have polyhedral morphology and exist in five modifications for the α- and β- forms while no spherulite structures have been observed for the γ-form. The three types of α-spherulites [238] are characterized by their cross-branched lamellae, while the β-spherulites exhibit leaflet-like lamellae [239–244]. In addition to its spherulitic form, iPP may also exhibit a lamellar or smectic morphology which is far less important as far as the filled PP is concerned [245]. This form is not thermodynamically stable and can be tranformed to a spherulitic one upon heating. Substantial variations in morphology can be observed within a single specimen due to different crystallization conditions in the core and skin regions, respectively, such as in injection molded parts. Injection molded PP articles, with the exception of very thin wall parts, exhibit a surface skin composed of lamellae and highly oriented long-chain molecules and an inner core composed of spherulites with no preferential molecular orientation [246–248]. It is worthy of note that the fast crystallization in the surface skin results in small and imperfect lamellae contributing to enhanced fracture resistance compared to the core. The thickness of the surface skin increases with increasing thermal conductivity of the polymer achieved commonly by adding inorganic fillers [249,250]. On the other hand, the skin thickness decreases with increasing molar weight [251]. Copolymerization with ethylene reduces skin thickness compared to the homopolymer under the same process conditions [247].

At very small strains within the viscoelastic region ($\varepsilon<0.5\%$), deformation within the PP solid is confined to disordered amorphous regions [97,98) due to their inherently low stiffness at temperatures above their T_g (−10 °C). In this deformation region, spherulites undergo affine deformation on the whole. Inside the spherulites, rotation of lamellae occurs [101]. The resulting orientation of lamellae depends on their position within the spherulite in respect to the orientation of the external deformation [102–104].

At larger strains, crystalline lamellae undergo plastic deformation processes manifested by macroscopic yielding. Lamellae under tension in their own planes fracture almost immediately with the microcracks being bridged by fine microfibrils formed by extended long-chain molecules. The complete fibrilation process can eventually consume the whole lamella when strained at extremely low strain rates and molecules are sufficiently long or there are stronger intermolecular attractions than weak van der Waals forces existing in common polyolefins [113]. A small-size spherulitic structure has more evenly distributed stresses resulting in a more even distribution of loads acting on individual chains. At very large deformations with draw ratios greater than 5, a highly oriented fibrilar structure is formed consisting of microfibrils clumped together in coarser fibrils bridged by unoriented microfibrils crossing longitudinal voids between them. Excellent treatment of this topic has been provided by Friedrich [107].

Polypropylene has recently become an attractive candidate for many engineering applications [190]. Relatively low price, excellent chemical resistance, good processability and the possibility of modifying its mechanical properties in a wide range by adding fillers and dispersions of secondary polymeric inclusions has contributed to its massive expansion into automotive, land transport, home appliances and other industries. Poor low-temperature impact behavior and relatively low stiffness are among the most important deficiences prohibiting neat PP replacing more expensive engineering thermoplastics in more demanding applications. Binary combinations of PP with fillers or elastomers address generally only one concern and exhibit either increased stiffness or enhanced low-temperature fracture resistance [187,188,252]. It is, however, necessary, in order to increase PP marketability into more demanding markets, to increase both stiffness and toughness at the same time. Hence, attempts have been made to incorporate both stiff inorganic filler and elastomer inclusions into the PP matrix in the course of melt mixing [252]. Unlike in the simple binary blends, phase morphology or, in other words, the relative spatial arrangement of filler and elastomer inclusions, plays a dominant role in determining the mechanical response of these complex materials [253–255].

Ternary composites of PP containing both rigid filler and elastomer have been shown to have the potential to be more rigid and tougher than neat PP itself [256,257]. The morphology of these ternary systems depends substantially on the rheological parameters of the polymer components and their relative surface free energies. The principal limits of the possible spatial arrangements, however, are determined by equilibrium thermodynamic considerations. As a result, two stable limiting morphologies may exist. First, a complete separation of filler and elastomer is achieved when there is a perfect adhesion between the filler and the PP matrix ($\gamma_f = \gamma_{PP} >> \gamma_e$). In the case of high surface energy inorganic fillers [$CaCO_3$, $Mg(OH)_2$, etc.], one can satisfy such a requirement by grafting PP with a polar co-monomer, such as maleic anhydride (MAH) or acrylic acid (AA). The second limiting case is that when a complete encapsulation of the filler by an elastomer layer exists. This can be done, on the other hand, by modifying the elastomer with a polar monomer such as MAH or AA. Due to the high viscosity of the PP melt, thermodynamic equilibrium is not achieved and a kinetic quasi-equilibrium state is established resulting in a degree of inclusion encapsulation strongly dependent on the mixing conditions. One has to keep in mind, however, that in this case the large shear stresses existing in the melt during its mixing will tend to de-wet or peel-off the elastomer from the filler surface and, hence, it has been shown that it is much harder to obtain this morphology experimentally compared to the case of complete separation [253].

9.1 Change in PP morphology Due to Presence of Solid Inclusions

There has been no report of any change in constitution (atomic structure) or configuration (primary bond spatial arrangement) of PP molecules, except of partial reduction of the average molecular weight due to its interaction with fillers. There have been reports in the literature revealing changes in conformation of long-chain molecules in the imminent vicinity of the surface of solid inclusions. In most reports, it has been suggested that chains are less densely packed in this interphase or, in other words, they exist as expanded coils [258]. In contrast, Vollenberg et al. [259] explained the effect of particle size on the elastic modulus of both glassy and semicrystalline polymers by the formation of an interphase of increased density in the vicinity of the surface of the inclusions. The morphology of the crystalline phase in terms of degree of crystallinity and lamellae thickness is affected substantially by the presence of fillers, including the effect of reduced molecular weight upon crystallization.

During solidification from the melt, gradients of mass transport occur near the filler surface caused by the potential field at the surface of the solid inclusion and fluctuation in local temperature [260]. As a result, both the supermolecular structure and the chain mobility of PP in a thin layer near the filler surface is altered compared to the bulk, with the extent of such a change proportional to the surface activity of the filler and the topology of the surface [261–268]. Surface affinity is usually defined as the ratio between adhesive and cohesive energies of the filler and matrix, and, generally, more surface active fillers cause greater changes in PP morphology. The thickness of such a layer ranges from 10 to several hundreds of nanometers [269] and has more or less an amorphous character.

In addition to the formation of such an interphase, fillers affect the morphology of PP in the bulk. Many fillers act as primary nucleation agents increasing the number of spherulites and reducing their mean diameter without any significant modification of the crystallization kinetics. A slight decrease in the degree of crystallinity by about 10% has been observed using X-ray diffraction measurements. Interpretation of these measurements is partly obscured by peak overlaps from the PP and the filler and by the reduced perfectness of the crystallites formed in the filled PP. Up to a particulate filler content of about 20 vol%, α-form lamellae are formed similar to unfilled PP. The thickness of the lamellae is also reported to be reduced in filled PP [39,40]. Above 20 vol% of the filler, the spherulite morphology of PP disintegrates into a system of lamellae oriented more or less randomly in respect to the filler surface.

9.2 Change in Physical Properties of PP by Fillers

Changes in the glass transition temperature, T_g, are a sensitive indicator of the structural changes occuring near the filler surface [270]. At constant filler loading, T_g increases with the increase in the average molecular mass of PP and, at constant molecular mass it increases with filler content [271]. The observed T_g shifts are relatively modest ranging for PP from 3 to 15 °C. The difference in T_g shifts between various fillers, at otherwise the same conditions, is ascribed to the different surface activity of the respective fillers.

Changes in other properties of PP, such as elastic moduli, yield strength, fracture toughness, etc. result from the changes in the PP morphology. For practical reasons (interpretation of data), however, the matrix properties are assumed unchanged when analyzing measurements performed on filled PP and evaluating effects of fillers. One can hypothesize that by measuring the properties of filled PP at a filler volume content less than 1 vol% and extrapolating these data to zero filler content, one could obtain a reasonable estimate of the properties of the PP matrix. Taking data from literature [272–275], and performing such an extrapolation, one can conclude that the elastic modulus of the PP matrix is unaffected by the filler, the yield strength is slightly reduced and the fracture toughness is slightly increased. These conclusions are in qualitative agreement with the predictions based on the observed change in morphology [273]. More detailed treatment of the effect of fillers on the mechanical properties of thermoplastics will be provided in volume II of this work.

10 References

1. Gruenwald G (1992) Plastics, how structure determines properties. Hanser, Munich
2. Rubin II (ed) (1990) Handbook of plastic materials and technology. J Wiley Interscience, New York
3. Utracki LA (1990) Polymer alloys and blends. Hanser, Munich, New York, p 1
4. Matonis VA, Small NC (1969) Polym Eng Sci 9:91
5. Jancar J, DiBenedetto AT, DiAnselmo A (1996) Chem Papers 50:1
6. Theocaris PS (1983) Mesophase concept in composite materials. Springer, Berlin Heidelberg New York, p 3
7. Goan JC, Pprosen S (1969) In: Interfaces in composites, ASTM STP 452, American Society for Testing and Materials, Philadelphia, p 3
8. Greszczuk LB (1969) in Interfaces in composites, ASTM STP 452, American Society for Testing and Materials, Philadelphia, p 42
9. DiAnselmo A, Jancar J, DiBenedetto AT, Kenny JM (1992) In: DiBenedetto AT, Nicolais L, Watanabe R (eds) Composite materials. Elsevier, London, p 49
10. Lipatov Yu S (1977) Physical chemistry of filled polymers. (Eng transl.) Int Polym Sci Tech Monograph No. 2, Moscow
11. He MY, Hutchinson JW (1989) Int J Solid Structure 25:1053
12. Mikata Y, Taya M (1985) J Appl Mech 52:19
13. Piggott MR (1987) Polym Compos 8:291
14. Tsai HC, Arocho AM, Gause LW (1990) Mater Sci Eng A126:295
15. Johannson OK, Stark FO, Vogel GE, Lacefield RM, Baney RH, Flaningam OL (1969) In: Interfaces in composites, ASTM STP 452, American Society for Testing and Materials, Philadelphia, p 168
16. Pukanszky B, Tudos F, Jancar J, Kolarik J (1989) J Mater Sci Lett 8:1040
17. Jancar J, DiBenedetto AT (1994) J Mater Sci 29:4651
18. Jancar J, DiAnselmo A, DiBenedetto AT (1992) Polym Eng Sci 32:1394

19. Hancock M Filled Thermoplastics. In: Rothon R (ed) Particulate filled polymer composites. Longman Scientific & Technical, Harlow (UK), p 279
20. Ward IM (1983) Mechanical properties of solid polymers, 2nd edn. John Wiley, New York
21. Keith HD (1986) Morphology of polymers. In: Bever MB (ed) Encyclopedia of materials science and engineering. Pergamon, New York, p 3110
22. Ferry JD (1980) Viscoelastic properties of polymers, 3rd edn. John Wiley, New York
23. Struik LCE (1990) Internal stresses, dimensional instabilities and molecular orientation in plastics. John Wiley, New York, p 47
24. Tryson LD, Kardos JL (1981) Proc, 36th. Ann Conf Reinforced Plastics/Composite Institute, SPI, Session 2-E
25. Andrews EH (1968) Fracture in polymers. Oliver and Boyd, London
26. Kinloch AJ, Young RJ (1983) Polymer fracture. Elsevier, London, p 147
27. Kausch HH (1978) Polymer fracture. Springer, Berlin Heidelberg New York, p 9
28. Billingham NC, Jenkins AD (1972) The principles and methods of polymer preparation. In: Jenkins AD(ed) Polymer science, a materials science handbook. North Holland Publ. Co., London, chap 1, pp 1–121
29. Kucera M (1992) Mechanism and kinetics of addition polymerizations. Academia Press, Prague
30. Monasse B, Haudin JM (1995) In: Karger-Kocis J (ed) Polypropylene: structure, blends and composites, vol 1. Chapman & Hall, London, p 5
31. Schnell H (1964) Chemistry and physics of polycarbonates. John Wiley Interscience, New York
32. Odian GB (1981) Principles of polymerization, 2nd edn. John Wiley, New York
33. Davies P, Plummer CJG (1993) Structure and mechanical properties of other advanced thermoplastic matrices and their composites. In: Kausch H-H (ed) Advanced thermoplastic composites: characterization, processing. Hanser, Munich
34. Eby K, Evers RC, Meador MA, Wilson D (eds) (1994) High performance polymers and polymer matrix composites. MRS Symp Proc, vol 305, Pittsburg
35. Brady RL, Porter RS (1990) J Appl Polym Sci 39:1873
36. Kardos JL, Cheng FS, Tolbert TL (1973) Polym Eng Sci 13:455
37. Jancar J, DiBenedetto AT (1993) J Mater Sci Mater Medicine 4:555
38. Magill JH (1977) In: Schultz JM (ed) Treatise on materials science and technology, vol 10. Academic Press, New York, p 3
39. Rybnikar F (1981) Macromol Sci-Phys B19:1
40. Rybnikar F (1982) J Appl Polym Sci 27:1479
41. Jancar J (1987) Deformation behavior of filled polypropylene. PhD thesis, Institute of Macromolecular Science, Czech Academy of Sciences, Prague
42. Bassett DC (1981) Principles of polymer morphology. Cambridge Univ. Press, New York, p 16
43. Isayev AI, Chan TW, Shimojo K, Gmerek M(1995) J Appl Polym Sci 55:807
44) Isayev AI, Chan TW, Shimojo K, Gmerek M (1995) J Appl. Polym. Sci 55:821
45. van der Mark JM, Kiel AM, Kolloid (1970) Z und Z. Polym. 237:236
46. McKenna GB, Statton WO (1976) J Polym Sci Macromol Rev 11:1
47. Dealy JM, Wissbrun KF (1989) Melt rheology and its role in plastics processing. Van Nostrand Reinhold, New York
48. Wunderlich B (1976) Crystal nucleation, growth, annealing. Academic Press, New York
49. Galeski A (1995) Nucleation of polypropylene. In: Karger-Kocsis J (ed) Polypropylene: structure, blends composites, vol 1. Chapman & Hall, London, p 116
50. Flory P (1962) J Am Chem Soc 84:2857
51. Binsbergen FL, Lange BGM (1970) Polymer 11:309
52. Binsbergen FL (1970) Koll Z und Z Polym 237:289
53. Birnsbergen FL (1970) Polymer 11:253
54. Birnsbergen FL (1972) J Crystal Growth 16:249
55. Birnsbergen FL (1973)J Polym. Sci Polym Phys Ed 11:117

56. Pospisil L, Jancar J, Rybnikar F (1990) J Mater Sci Lett 9:495
57. Hoffman JD, Davies GT, Lauritzen JI Jr (1976) The rate of crystallization of linear polymers. In: Hannay NB (ed) Treatise on Solid State Chemistry, vol 3. Plenum Press, New York, p 497
58. Hoffman JD (1982) Polymer 23:565
59. deGennes P-G (1971) J Chem Phys 55:572
60. Hoffman JD, Gutman CM DiMarzio EA (1979) Discussions of the Faraday Soc 68:177
61. DiMarzio EA, Gutman CM Hoffman JD (1979) Discussions of the Faraday Soc 68:210
62. Hoffman JD, Miller RL (1988) Macromolecules 21:3038
63. Clark EJ, Hoffman JD (1984) Macromolecules 17:878
64. Khoury F, Passaglia E (1976) In: Hannay NB (ed) Treatise on solid state chemistry, vol 3. Plenum Press, New York, p 335
65. Geil PH (1963) Polymer single crystals. John Wiley Interscience, New York
66. Mandelkern L (1989) Crystallization and melting of polymers In: G Allen (ed) Comprehensive polymer science, vol 2. Pergamon Press, Oxford
67. Avrami M (1939) J Chem Phys 7:1103
68. Avrami M (1940) J Chem Phys 8:812
69. Avrami M (1941) J Chem Phys 9:1977
70. Tobin MC (1974) J.Polym Sci Polym Phys 12, 399
71. Tobin MC (1976) J Polym Sci Polym Phys 14:2253
72. Nakamura K, Katayama K, Amano T (1975) J Appl Polym Sci 17:1031
73. Ozawa T (1971) Polymer 12:150
74. Schneider W, Koppl A, Berger J (1988) Intern Polym Proc 2:151
75. Eder G, Janeschitz-Kriegl H (1995) In: Meijer HEH (ed) Materials science and technology, vol 18. Verlag Chemie, Weinheim
76. Eder G, Janeschitz-Kriegl H, Liedauer S (1990) Prog Polym Sci 15:629
77. Miller RL (1966) In: Brandrup J, Immergut EH (eds) Polymer handbook. Interscience, New York, chap 3, p 1
78. Kakudo M, Kasai U (1972) In: X-ray diffrection by polymers. Elsevier, New York
79. Hosemann R (1972) Crit Rev in Macromol Sci 1:351
80. Keith HD (1964) J Polym Sci 2A:4339
81. Keith HD, Padden JF, Jr (1963) J Appl Phys 34:2409
82. Keith HD, Padden JF, Jr (1964) J Appl Phys 35:1270
83. Keith HD, Padden JF, Jr (1984) Polymer 25:28
84. Keith HD, Padden JF, Jr (1959) JPolym Sci 41:525
85. Magill JH (1977) In: Schultz JM (ed) Treatise on materials science and technology, vol 10. Academic Press, New York, p 168
86. Ballard DGH, Wignall GD, Schelten J (1973) Eur Polym J 9:965
87. Fischer EW, Wendorff JH, Dettenmaier M, Lieser G, Voigt-Martin I (1974) Amer Chem Soc Polym Preprints 15:8
88. Kirste RG, Kruse WA, Schelten J (1972) Macromolecular Chem 162:299
89. Yeh GSY (1972) CRC Crit Rev Macromol Sci (213) April
90. Geil PH (1974) Amer Chem Soc Polym Preprints 15:35
91. Lindenmeyer PH (1974) In: Geil PH, Baer E, Wade Y, (eds) The solid state of polymers. Dekker, New York, p 361
92. Williams JG (1984) Polymer fracture. Ellis Horwood, Chichester
93. Voight-Martin IG, Fischer EW, Mandelkern L (1980) J Polym Sci Polym Phys 18:2347
94. Samuels RJ (1985) Polym Eng Sci 25:864
95. Barham PJ, Keller A (1977) J Mater Sci 12:2141
96. Billmeyer FW, Jr (1983) Textbook of polymer science, 3rd edn. John Wiley, New York
97. Hartman B, Lee G, Wong W (1987) Polym Eng Sci 27:823
98. Samuels RJ (1979) J Polym Sci Phys 17:535
99. Halpin JC, Kardos JL (1972) J Appl Phys 43:2235
100. Iobst SA, Manson JA (1972) J Polym Sci A1-10:179

101. Kataiama T et al. (1974) J Macromol Sci Phys B9:35
102. Boyd RH, Aylwin PA (1984) Polymer 25:323
103. Boyd RH, Aylwin PA (1984) ibid 330
104. Boyd RH, Aylwin PA (1984) ibid, 340
105. Hay IL, Keller A (1965) Kolloid Z 204:43
106. Peterlin A (1975) In: Ward IM (ed) Molecular aspects of oriented polymers. John Wiley, New York, p 36
107. Friedrich K (1978) Uber den Einfluss der Morphologie auf Festigkeit und Bruchvorgange in isotaktischem Polypropylen, PhD dissertation, University of Bochum, Germany
108. Shames ICH (1965) Mechanics of deformable solids, 2nd edn. Prentice-Hall, New York, p 99
109. Bueche F (1952) J Chem Phys 20:1959
Bueche F (1954) J Chem Phys 22:603
Bueche FJ (1955) J Appl Phys 26:738
Bueche F (1956) J Chem Phys 25:599
110. Rouse PE (1953) J Chem Phys 21:1272
111. Zimm BH (1950) J Chem Phys 18:830
Zimm BH (1956) J Chem Phys 24:269
112. Tobolski AV (1960) Properties and structure of polymers, John Wiley, New York, p 166
113. Kolarik J (1987) In: High modulus polymer fibers and fibrous composites, Academia Press (Czech), Prague, p 10
114. Andrews EH (1973) In: Haward RN (ed) Physics of glassy polymers. Applied Science Publ, London, p 394
115. Dijkstra K (1993) Deformation behavior of toughened Nylon 6, PhD dissertation, University of Twente
116. Bowden PB (1973) In: Haward RN (ed) The physics of glassy polymers. Applied Science Publ, London, p 279
117. Bowden PB, Saha S (1974) Phil Mag 29:149
118. Bowden PB, Saha S (1970) Phil Mag 22:463
119. Bowden PB (1970) Phil Mag 22:455
120. Argon AS (1973) Phil Mag 28:839
121. Brown N (1983) J Mater Sci 18:2241
122. Kramer EJ (1975) J Polym Sci Polym Phys 13:509
123. Bowden PB, Young RJ (1974) J Mater Sci 9:2034
124. Lin TM, Harrison IM (1988) Polym Mater Sci 59:430
125. Gent AN, Madon S (1989) Polym Sci 12:1653
128. Bauwens-Crowet C, Ots JC, Bauwens JC (1974) J Mater Sci 9:1197
129. Haaf F, Breuer H, Stabenow J (1977) J. Macromol. Sci. Phys. B14: 387
130. Andrews RD, Kazama Y (1967) J Appl Phys 38:4118
131. Brady TE, Yeh GSY (1971) J Appl Phys 42:4622
132. Robertson RE (1968) Appl Polym Symp 7:201
133. Kambour RP, Robertson RE (1972) In: Jenkins AD (ed) Polymer science. North Holland, London, p 688
134. Thierry A, Oxborough RJ, Bowden PB (1974) Phil Mag 30:527
135. Argon AS, Bessonov MI (1977) Polym Eng Sci 17:174
136. Argon AS, Bessonov MI (1977) Phil MaG 35:917
137. Yamini S, Young RJ (1980) J Mat Sci 15:1814
138. Keller A, (1968) Rep Prog Phys 31:623
139. Peterlin A (1975) Mechanisms of deformation in polymeric solids. In: Polymeric Materials. American Society for Metals, Philadelphia, p 208
140. Young RJ (1979) In: Andrews EH (ed) Developments in polymer fracture. Applied Science Publ, London, p 223
141. Young RJ (1974) Phil Mag 30:85
142. Petermann J, Gleiter H (1973) J Mater Sci 8:673

143. Haward RN (1973) In: Physics of glassy polymers, Applied Science Publ, London, p 340
144. Friedrich K, Schafer JK (1979) J Mater Sci 14:480
145. Chau CC, Li JCM (1979) J Mater Sci 14:1593
146. Chau CC, Li JCM (1979) J Mater Sci 14:2172
147. Jang BZ, Uhlmann DR, van der Sande JB (1984) J Appl Polym Sci 29:3409; ibid 30:2485
148. Kramer EJ (1979) In: Andrews EH (ed) Developments in polymer fracture-1. Applied Science Publ, London, p 55
149. Henkee CS, Kramer EJ (1989) J Polym Sci Polym Phys 22:721
150. Berger LL, Kramer EJ (1988) J Mater Sci 23:3536
151. Argon AS (1975) Pure and Appl Chem 43:247
152. Argon AS, Hannoosh JG (1977) Phil Mag 36:1195
153. Haward RN, Owen DRJ (1973) J Mater Sci 8:1136
154. Gent AN (1973) J Macromol Sci B8:597
155. Sternstein SS, Ongchin L (1969) ACS Polym Prepr. 19(2):1117
156. Bowden PB, Oxborough RJ (1973) Phil Mag 28:547
157. Argon AS, Salama MM (1977) Phil Mag 35:1217
158. Donald AM, Kramer EJ (1981) Phil Mag A43:857
159. Williams JG, Marshall GP (1975) Proc RSoc A342:55
160. Marshall GP, Coutts LH, Williams JG (1974) J Mater Sci 9:1409
161. Haward RN, Daniels HE, Treolar LRG (1978) J Polym Sci Polym Phys 16:1169
162. Pitman GL, Ward IM, Duckett RL (1978) J Mater Sci 13:2092
163. Pitman GL, Ward IM (1979) Polymer 20:895
164. Kramer EJ (1979) J Mater Sci 14:1381
165. Farrar NR, Kramer EJ (1981) Polymer 22:691
166. Robertson RE (1976) In: Deanin RD, Crugnola AM (eds) Toughness and brittleness of plastics. ACS, Washington, p 89
167. Donald AM, Kramer EJ (1982) J Polym Sci Polym Phys 20:899
168. Donald AM, Kramer EJ, Bubeck RA (1982) J Polym Sci Polym Phys 20:1129
169. Kramer EJ (1975) J Polym Sci Polym Phys 13:509
170. Wu JC B, Li JCM (1976) J Mater Sci 11:434
171. Voigt-Martin I H, Wendorff J (1985) Amorphous polymers. In: Encyclopedia of polymer science and engineering, vol 1. John Wiley, New York, p 789
172. Chau CC (1981) Li JCM J Mater Sci 16:1858
173. Camwell L, Hull D (1973) Phil Mag 27:1135
174. Mills NJ (1976) J Mater Sci 11:363
175. Narisawa I, Ishikawa M, Ogawa H (1980) J Mater Sci 15:2059
176. Donald AM, Kramer EJ (1982) J Mater Sci 17:1871
177. Donald AM, Kramer EJ, Kambour RP (1982) J Mater Sci 17:1739
178. Bucknall CB, Clayton D, Keast WE (1972) J Mater Sci 7:1443
179. Cornes PL Haward RN (1974) Polymer 15:149
180. Cornes PL, Smith K, Haward RN (1977) J Polym Sci Polym Phys 15:955
181. Walker N, Haward RN, Hay JN (1979) J Mater Sci 14:1085
182. Walker N, Hay JN, Haward RN (1980) J Mater Sci 15:1059
183. Walker N, Haward RN, Hay JN (1981) J Mater Sci 16:817
184. Walker N, Hay JN, Haward RN (1979) Polymer 20:1056
185. Broutman L J, Panizza G (1971) Int J Polym Mater 1:95
186. Robertson RE (1969) J Polym Sci A-2:1315
187. Jancar J, DiBenedetto AT, DiAnselmo A (1993) Polym Eng Sci 33:559
188. Jancar J, DiAnselmo A, DiBenedetto AT, Kucera J (1993) Polymer 34:1684
189. Riew CK, Kinloch AJ (eds) (1993) Toughened Plastics I. Advances in Chemistry Series, ACS, Washington
190. Karger-Kocsis J (ed) (1995) Polypropylene: structure, blends and composites, vol 1. Chapman & Hall, London

191. Rothon RED (1995) In: Particulate filled polymer composites Longman Scientific & Technical, Harlow, UK
192. Goodier JN (1933) Trans Am. Soc Mech Engs 55:39
193. Oxborough RJ, Bowden PB (1974) Phil Mag 30:171
194. Guild FJ, Young RJ (1989) J Mater Sci 24:298
195. Guild FJ, Young RJ (1989) J Mater Sci 24:2454
196. DiAnselmo A, Jancar J, DiBenedetto AT, Kenny JM (1992) In: DiBenedetto AT, Watanabe R (eds) Composite materials. Elsevier, Amsterdam, p 49
197. Gaylord NG, Mark HF, Bikales NM (1970) Encyclopedia of polymer science and technology, vol 11. Wiley Interscience, New York, p 597
198. Monasse B, Haudin JM (1995) Molecular structure of polypropylene homo- and copolymers. In: Karger-Kocsis J (ed) Polypropylene: structure, blends and composites, vol 1. Chapman & Hall, London, p 5
199. Monasse B, Haudin JM (1988) Colloid and Polym Sci 266:679
200. Schultz J-M (1984) Polym Eng Sci 24:770
201. Kardos J, Christiansen AW, Baer EA (1966) J Polym Sci A2:777
202. Poe KD (1968) J Polym Sci A2:657
203. Awaya H (1966) J Polym Sci B4:127
204. Plazek DL, Plazek DJ (1983) Macromolecules 16:1469
205. Turner-Jones A, Aizlewood JM, Beckett DR (1964) Makromol Chem 75:134
206. Lotz, B, Lovinger AJ, Cais RS (1988) Macromolecules 21:2375
207. Monasse B, Haudin JM (1985) Coll. Polym Sci 263:822
208. Martuscelli E, Pracella M, Zambelli A (1980) J Polym Sci Polym Phys Ed 18:619
209. Mucha M (1981) J Polym Sci Polym Symp 69:79
210. Samuels RJ (1975) J Polym Sci Polym Phys 13:1417
211. Cheng SZD, Janimak JJ, Rodriguez J (1995) Crystalline Structures of polypropylene homo- and copolymers. In: Karger-Kocsis J (ed) Polypropylene: structure, blends and composites, vol 1. Chapman & Hall, London, p 33
212. Varga J (1995) Crystallization, melting and supermolecular structure of isotactic PP. In: Karger-Kocsis J (ed) Polypropylene: structure, Blends and composites, vol 1. Chapman & Hall, London, p 56
213. Ewen JA, Johns RL, Razavi A, Ferrara JD (1988) J Am Chem Soc 110:6255
214. Chatani Y., Maruyama H (1991) J Polym Sci Polym Phys 29:1649
215. Chatani Y, Maruyama H, Noguchi K (1990) J Polym Sci Polym Lett. 28:393
216. Lovinger AJ, Davis D, Lotz B (1991) Macromolecules 24:552
217. Lovinger AJ, Lotz B, Davis D, Padden FJ, Jr (1993) Macromolecules 26:3494
218. Lovinger AJ, Lotz B, Davis D (1990) Polymer 31:2253
219. Rodriguez-Arnold J, Zhang A, Che ZD, Lovinger AJ (1994) Polymer 35:1884
220. Wunderlich B (1973) Macromolecular physics, vol 1. Academic Press, New York
221. Varga J (1992) J Mater Sci 27:2557
222. Fox D, Labes MM, Weissberger A (1965) Physics and chemistry of organic solid state, vol 1. Wiley Interscience, New York, chap 8
223. Vaughan AS, Bassett DC (1989) Crystallization and morphology. In: Allen G, Bevington JC (eds) Comprehensive polymer science, vol 1. Pergamon Press, London, chap 12
224. Peterlin A (1975) Molecular aspects of oriented polymers. In: Ward JW (ed) Structure and properties of oriented polymers. Elsevier, London, p 46
225. Keller A (1977) Routes to high modulus by ultraorientation of flexible molecules. In: Cifferi A, Ward JM (eds) Ultra-high modulus polymers. Elsevier, London, chap 11
226. Gray DG (1974) Polymer Letters 12:645
227. Thomason JL, van Rooyen AA (1992) J Mater Sci 27:897
228. Jenckel E, Teege E, Hinrichts W (1952) Koll Z 129:19
229. Fujiyama M, Wakino T (1991) J Appl Polym Sci 42:9
230. Xavier SF, Sharma YN (1984) Angewandte Makromol Chem 127:145
231. Garton A, Kim S W, Wiles DM (1982) J Polym Sci Polym Lett 20:273

232. Devaux E, Chabert B (1990) Polym Commun 31:391
233. Keith HD, Padden JF, Jr (1963) J Appl Phys 34:2400
234. Keith HD, Padden JF Jr (1986) Polymer 27:1463
235. Bassett DC, Vaughan AS (1986) Polymer 27:1472
236. PiorkowskAE, Galeski A (1985) J Polym Sci Polym Phys 23:1723
237. Schulze GEW, Willers R (1987) J Polym Sci Polym Phys 25:1723
238. Padden JF, Keith HD (1959) J Appl Phys 30:1479
239. Hendra PJ et al. (1984) Polymer 25:785
240. Struick LCE (1982) Plast Rubb Proc Appl 2:41
241. Samuels RJ (1967) J Polym Sci Part C 20:253
242. Samuels RJ, Yee RJ (1972) J Polym Sci Polym Phys A2, 10:385
243. Norton DR, Keller A (1985) Polymer 26:704
244. Idrissi M B O., Chabert B, Guillet J (1985) Makromol Chem 186:881
245. Peterlin A (1977) J Appl Phys 48:4099
246. Fujiyama M, Wakino T (1988) J Appl Polym Sci 35:29
247. Fujiyama M (1995) Higher order structure of injection molded polypropylene. In: Karger- Kocsis J (ed) Polypropylene: structure, blends and composites, vol 1. Chapman & Hall, London, chap 6, p 168
248. Janeschitz-Kriegel H, Fleischman E, Geymayer W (1995) Processing-induced structure formation. In: Karger-Kocsis J (ed) Polypropylene: structure, blends and composites, vol 1. Chapman & Hall, London, chap 10, p 295
249. Fujiyama M, Wakino T (1991) J Appl Polym Sci 43:97
250. Fujiyama M, Wakino T (1991) J Appl Polym Sci 42:9
251. Fyjiyama M, Wakino T (1991) J Appl Polym Sci 43:57
252. Kolarik J, Lednicky F, Jancar J, Pukanszky B (1990) Polym Commun 31:201
253. Kolarik J, Jancar J (1992) Polymer 33:4961
254. Jancar J, DiBenedetto AT (1993) Proc ANTEC'93, SPE 1993, vol II, pp 1698–1700
255. Jancar J, DiBenedetto AT (1993) In: Kahovec J (ed) Macromolecules 1992. VSP Publishers, pp 399–409
256. Jancar J, DiBenedetto AT (1995) J Mater Sci 30:1601
257. Jancar J, DiBenedetto AT (1995) J Mater Sci 30:2438
258. Jancar J, DiBenedetto AT (1994) Sci Eng Compos Mater 4:217
259. Vollenberg P H.T, Heikkens D (1986) Polymer 30:1656
260. Lipatov Yu S, Lebedev EV (1979) Macromol Chem Suppl 2:51
261. Fitchmon DR, Newman S (1970) J Polym Sci A28:1545
262. Johnson OK (1967) J Compos Mater 1:278
263. Kaas RL, Kardos J (1971) Polym Eng Sci 11:11
264. Kardos J et al. (1973) Polym Eng Sci 13:455
265. Morgan RJ (1974) J Mater Sci 9:1219
266. Pukanszky B, Tudos F, Jancar J, Kolarik J (1989) J Mater Sci Lett 8:1040
267. Pukanszky B, Turczanyi B, Tudos F (1988) In: Ishida H (ed) Interfaces in polymer, ceramic and metal matrix composites. Elsevier, New York, p 467
268. Jancar J, Vesely P, Kucera J (1991) J Mater Sci 26:4878
269. Ishida H (1984) Polym Compos 5:101
270. Chow TS (1984) Polym Eng Sci 24:1079 (1984)
271. Paakonnen EJ, Magonov SN, Tormala P (1986) In: Sedlacek B (ed) Polymer composites. deGruyter, Berlin, p 183
272. Jancar J (1998) "Mechanical properties of thermoplastic composites with engineered interphases" in "Polypropylene Handbook", H. Karian Ed., M. Dekker, New York 1998
273. Tong SN, Chen ML (1991) Analysis of transition temperatures in polymer-filler systems. In: Mitchell J (ed) Applied polymer analysis and characterization, vol II. Hanser, Munich, chap 5, p 329

Received: March 1998

Mineral Fillers in Thermoplastics: Filler Manufacture and Characterisation

R.N. Rothon

Department of Materials Technology, Manchester Metropolitan University, 3 Orchard Croft, Guilden Sutton, Chester, CH3 7SL, United Kingdom
Fax: + 44 (0) 1244 300375

The principal mineral fillers used in thermoplastics and the reasons for using them are identified, together with those features that have to be controlled in order to achieve the optimum results and to avoid associated deleterious effects. General methods of filler production are outlined in the light of these requirements and their application to the fillers in most use is described. Attention is given to the use of surface modification methods where these are part of the production process.
Considerable attention is also given to the methods that need to be used to adequately characterise a filler for use in thermoplastic applications, particularly the determination of particle shape and size.

Keywords: Fillers, Manufacture, Characterisation, Calcium carbonate, Talc, Mica, Wollastonite, Clays, Aluminium hydroxide, Magnesium hydroxide

1
Introduction

First introduced industrially in the 1930s, thermoplastic polymers are today produced and consumed in vast quantities and play a major role in many aspects of our everyday lives. It is estimated that over 16 million tons were consumed in Western Europe alone in 1991 [1]. Mineral fillers have, since the beginning, made an important contribution to the spectacular growth of thermoplastic polymers. The addition of mineral materials was initially seen mainly as a means of "extending" or reducing the compound cost but, as the relative cost of the polymers decreased, this became less important and attention was more and more focused on the property improvements that could be achieved.

It is estimated that over one million tons of mineral fillers were used in thermoplastic applications in western Europe in 1986 [2], and the figure is doubtless much greater today. Mineral fillers are used to some extent in virtually all the commercially important thermoplastic polymers but, in volume terms, the principal markets are in PVC and polyolefins, where calcium carbonate dominates the filler types with over 80% of the volume consumption [2].

Although there is an extensive scientific literature dealing with the complex subject of the effects of mineral fillers on the properties of thermoplastics, there is much less dealing with the equally important area of filler production, much of the information in this area being regarded by producers as proprietary. Useful information on some aspects can be found in a number of works [3–8], but much of what is published is in trade rather than scientific literature and, although some of this is of a high standard, it is not readily accessible.

This review analyses the role that mineral fillers play in today's thermoplastics and identifies the filler characteristics that are needed to meet the market needs successfully. The origin and methods of production of the principal commercial products is then described. Attention is also given to the area of filler characterisation, as a detailed understanding of the scope and limitations of some of the techniques and terminology in common practice is essential in understanding much of the work on structure/property relationships that will be encountered in the other parts of this work and in the scientific and trade literature in general. The important subject of the use of filler surface modifiers is only covered here in the context of production and characterisation methods as the basic science is covered in detail later in this work. The term "mineral fillers" has been interpreted fairly broadly and taken as including any particulate inorganic material. This allows the inclusion of important synthetic products of ultimate mineral origin, including short glass fibre.

2
The Effects of Mineral Fillers in Thermoplastic Polymers

2.1
General

The importance of the use of mineral fillers to the growth of applications for thermoplastic polymers has already been described. The addition of such materials affects most of the significant properties of the matrix, some beneficially, others detrimentally. Only some of these altered properties are important to the use of thermoplastics, and an appreciation of what these are is critical to identifying those filler characteristics that are important and in understanding how certain filler types and production methods have come to dominate the market.

The principal limitations of bulk thermoplastics that the use of mineral fillers is targeted at overcoming are cost, shrinkage on moulding, stiffness and flammability. Important properties that must not be too adversely affected by the fillers are cost, appearance, toughness, density and ageing performance. There are also special uses in products such as polymer film. The need to optimise the benefits of filler use against the potential disadvantages is a consistent theme in the industries involved in filler production and use, and largely determines the characteristics of the products in commercial use. These important properties are discussed in more detail below, with the emphasis being on their implications for filler choice and the production method to be employed.

2.2
Cost

The need to minimise production costs and hence the selling price of mineral fillers is a dominant theme running through the technology. Although compound cost reduction was one of the principal reasons for using fillers in the past, this is of less importance now, due to the fall in the cost of the commodity thermoplastics. Indeed, the cost in use of fillers (taking into account volume costs and the cost of incorporation) can now lead to a rise in overall compound costs and it is often the minimisation of this cost increase that has become important.

In calculating the effect of fillers on costs one must remember that polymers are generally used by volume, while both filler and polymer costs are usually quoted by weight. Most mineral fillers are considerably denser than polymers (usually 2–3 times) and hence their effective cost is considerably higher than appears at first sight. Some idea of the equivalent volume cost for fillers in the main thermoplastic polymers is given in Table 1, using estimated 1996 European price levels.

In addition to the raw material cost, one also has to take into account compounding costs, the cost of any coupling agent or surface modifier that is not already present on the filler and the cost of additional stabilisers that might be re-

Table 1. Approximate relative volume costs of common mineral fillers and thermoplastic polymers

Filler or polymer type	Approximate relative volume cost (polyethylene=1.0)
Ground calcium carbonates	0.2–0.5
Clays	0.3–0.5
Calcined clays	1.4
Coated calcined clays	2.4
Coated precipitated calcium carbonates	1.9
Wollastonite (uncoated)	1.5
Talcs	1.0–1.3
PVC	1.5
Polyethylene	1.0
Polypropylene	1.0
Polyamides	3.0–6.0

quired. In some instances, these extra costs can be quite considerable. Where polymers, notably PVC, are already compounded in order to incorporate various solid additives, then the additional cost of incorporating a filler is considerably lower than those, such as polyolefins, where such a step is not commonly used and extra equipment is needed. Not surprisingly the use of fillers is more widespread in the former case.

2.3 Processing

Mineral fillers have a marked influence on the compounding procedures used for thermoplastics processing. These must be chosen to ensure good dispersion, while producing minimum aspect ratio degradation when fibre or plate fillers are being used.

The choice of processing equipment is covered later in this work, but some comments are appropriate here concerning filler effects. In general fillers raise the melt viscosity and this can be a serious problem, especially in extrusion processing. The amount by which this occurs is dependent on the particle shape and size distribution and on the filler surface properties. These parameters are usually controlled during filler processing to obtain the lowest viscosities commensurate with the desired end effect (e.g. some viscosity increase has to be tolerated when high aspect ratio fillers are being used for increased stiffness).

Filler surface treatments, such as fatty acids, are very useful for reducing melt viscosities and some fillers would be impossible to use at the loadings needed for certain applications such as fire retardancy without some form of surface treatment. In some cases melt viscosities can be maintained at similar levels to the unfilled polymer, even in highly filled systems, despite the use of high filler loading [9].

2.4 Shrinkage on Moulding

Because of their semi-crystalline nature, thermoplastics exhibit a considerable shrinkage on cooling from the melt and this can cause problems in achieving accurate mouldings. The presence of mineral fillers reduces the overall level of shrinkage, but unless the particles are isotropic they can cause differential shrinkage leading to warpage problems [10].

One of the principal reasons for the use of mineral fillers in these polymers is to reduce this shrinkage problem. Platy fillers have generally been found to be the most effective in this respect, giving the highest levels of shrinkage reduction consistent with maintaining adequate properties in other areas.

2.5 Stiffness and Heat Distortion Temperature

One of the principal limitations of thermoplastic polymers is their low stiffness (modulus), especially at elevated temperatures (heat distortion). Certain filler types are very good at improving these properties and this is another of the main reasons why fillers are used in these polymers.

A great deal of work has been done in this area and, although much of it is of an empirical nature, the main factors controlling the effects of fillers is now reasonably well understood [11, 12]. It appears that modulus increases with filler loading and that high aspect ratio fillers (plates or fibres) give the best performance, with talcs, clays, wollastonite, micas and short glass fibres being widely used. There is some evidence that particle size is also important, with the stiffness increasing significantly when small particles are used [13, 14]. This is thought to be due to the formation of an immobilised polymer shell around the filler which is contributing to the overall stiffness and the amount of which becomes significant as the particle size reduces. The effect of surface chemistry is less clear with reports on the use of coatings ranging from little or no effect [15–17] to large positive or negative effects [18–20]. It seems that the effect of surface chemistry is very system specific.

It is important that the composite retains as much toughness as possible and, as will be described later, this can lead to important particle size restrictions. Toughness also seems to differ in its response to the different ways of increasing modulus (loading, aspect ratio, particle size) and this affects the optimum filler characteristics. Coating technology can be important in getting the best balance between toughness and stiffness.

2.6 Flammability

Another limitation of thermoplastics is their inherent flammability. Where this is important, it has traditionally been overcome by the use of very efficient halogen/antimony oxide combinations [21]. These traditional flame retardants have certain limitations, notably the formation of considerable amounts of smoke and corrosive gasses on combustion and some possible toxicity concerns [22]. As a result, a number of alternative technologies have been developed and are of growing importance.

One of the emerging technologies that is showing great promise is the use of hydrated mineral fillers such as aluminium and magnesium hydroxides, as such materials can provide high levels of flame retardancy without the formation of smoke or corrosive and potentially toxic fumes. The use of fillers as flame retardants has recently been reviewed by Rothon [23]. Essentially the key features are an endothermic decomposition to reduce the temperature, the release of an inert gas to dilute the combustion gases and the formation of an oxide layer to insulate the polymer and to trap and oxidise soot precursors.

Flame retardant fillers have been used for many years in thermosets, but have only recently begun to make an impression in the thermoplastics area due to the need to develop products sufficiently stable to withstand the higher processing temperatures involved.

The key requirements of fillers for this application are thus:

- stability at the polymer processing temperature;
- a highly endothermic decomposition at, or near to, the temperature of polymer degradation releasing large amounts of inert gasses (water, carbon dioxide);
- a voluminous and tenacious, high surface area oxide residue (ash).

Particle size and shape do not appear to be too important to the flame retardancy, although a relatively small particle size appears to be preferred [24]. There is, however, some evidence that the presence of a surface modifier can affect the flame retardancy [25] and this must be born in mind.

To be effective, these fillers have to be used at high loadings and it is essential to minimise any associated loss in important properties such as toughness. It is this aspect that largely determines the optimum particle morphology, rather than the flame retardancy.

In addition to the fire retardant fillers which are effective in their own right, a number of mineral fillers are used as components of fire retardant systems for thermoplastics. The principal one is antimony oxide.

2.7 Appearance

The addition of fillers to thermoplastic polymers usually affects their appearance. While not absolutely essential, it is generally regarded as beneficial if the filler produces a pleasant white appearance or has no effect on the colour. It is a distinct disadvantage if the filler produces dark colours and makes the composite harder to pigment.

The effect of particulate additives on the appearance of polymers is a complex subject depending, among other things, on the colour, refractive index, particle size and impurities of the particulate phase. Some excellent reviews of the physical laws involved can be found in publications from pigment producers [26]. In general the fillers in common use have little colour themselves, and as their refractive index is similar to that of the polymers, they have little pigmentary effect. Most mineral fillers do have anisotropic crystal structures and hence more than one refractive index and this can give rise to complex birefringent effects [27]. The main appearance problems are probably due to the presence of impurities, which can cause colour effects of their own or from a promotion of polymer degradation.

The main requirement in this area is thus to minimise the presence of deleterious impurities, notably transition elements and organic materials. This can be achieved by careful choice of raw materials and by avoiding contamination during processing, particularly milling. In some cases special purification steps are incorporated into the production process to remove problem contaminants.

An issue that has been receiving increasing attention is the deleterious effect of fillers on the scratch resistance of polymers, as measured by the loss in surface appearance. The understanding of this problem is still at a rudimentary stage, but it appears that the problem can be minimised by control of particle morphology [28] and correct choice of surface treatments [29].

2.8 Ageing

Some thermoplastics, notably polypropylene, are relatively unstable to heat and light unless stabiliser additives are present [30]. In the early days, these additives comprised a significant proportion of compound raw material costs, but over time the polymers and the stabiliser systems have been improved until they have only a relatively small effect on the costs of unfilled polymers today.

Mineral fillers can produce both pro-degradant and stabilising effects in thermoplastic polymers. The reasons for this are varied and complex with factors such as adsorption of stabilisers and the presence of detrimental trace impurities such as certain metals playing a part. Until recently this was a little explored area, but several useful reviews have recently appeared [31, 32].

2.9 Toughness

In most, but not all, cases the incorporation of mineral fillers at the levels needed to produce useful effects causes a significant decrease in the toughness or impact strength of the composite. The major exception is in polypropylene, where certain types of calcium carbonate can give a very significant increase in toughness at loadings of 30–40 wt% [33].

Minimising this loss in toughness is a key feature in much development work and this subject has recently been reviewed by [34]. The subject is a very complex one and despite all the effort to date is still far from completely understood. The factors involved have been identified as including filler size and shape distribution, filler surface treatment, polymer type and even the test used to measure the toughness.

With particle size, it is the distribution rather than the average size that is important and it has been reported that small amounts of oversized particles can cause a significant reduction in toughness [35, 36]. Too many fine particles (eg below 0.5 micron) can also be a problem due to the immobilised polymer effect [37]. In general it seems that a particle size distribution in the range 0.5–10 microns is optimal. The exact distribution within this range is still important in determining the packing behaviour of the filler in the polymer and hence the maximum loading that can be obtained before particle/particle interactions start to dominate and reduce the toughness.

It is generally held that toughness is adversely affected by increasing aspect ratio [38] although some work disputes this for specific systems [39].

Filler surface chemistry is clearly important, although the effects vary widely according to the exact nature of the filler, polymer and surface modifier. Some of the factors that can influence toughness and are, at least in part, controlled by filler surface chemistry include: the level of filler polymer interaction [40], the structure of heterophasic polymers [41], the amount of polymer degradation during compounding [42], filler dispersion [43] and polymer crystallinity arising from altered nucleation processes [44].

3
General Methods of Filler Production

3.1
Overview

The methods of filler production from mineral sources can be broadly classified into direct and synthetic methods. Direct methods are the cheaper of the two, but generally result in a less pure product and in less optimal particle size and shape. It is also believed that they can result in a less consistent product than synthetic processes, although most reputable producers have very high quality control standards. Direct methods tend to dominate the industry because of the need to minimise cost and because they can often produce a material that is more than adequate for the purpose. Synthetic routes are more suited to the production of high purity and ultrafine particles. Although these products are generally restricted to the low volume speciality ends of the thermoplastics market, their relatively high prices gives them an important position in terms of value.

Calcium carbonate fillers are produced by both direct and synthetic methods and this results in distinctly different morphologies as shown in Figs. 1 and 2.

Filler surface treatments are of considerable use in thermoplastic applications and are often applied as part of the production process. The various methods used to apply such treatments are therefore covered in this article.

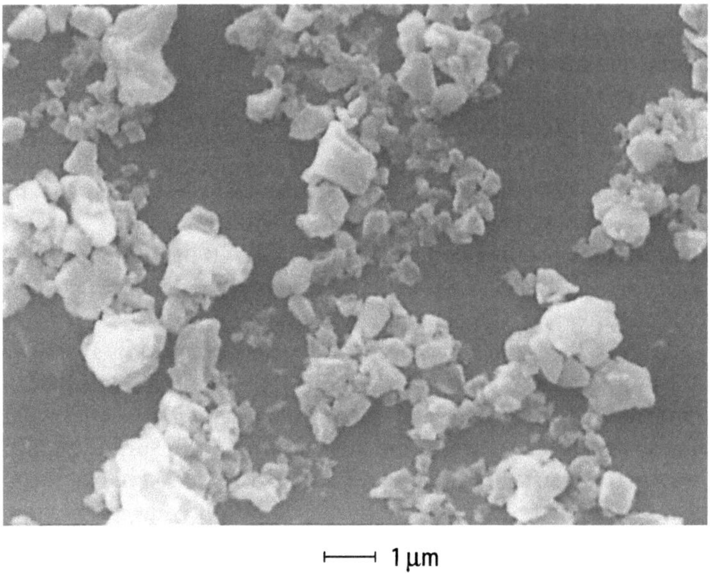

Fig. 1. Scanning electron micrograph of a typical ground calcium carbonate filler (courtesy of ECC International)

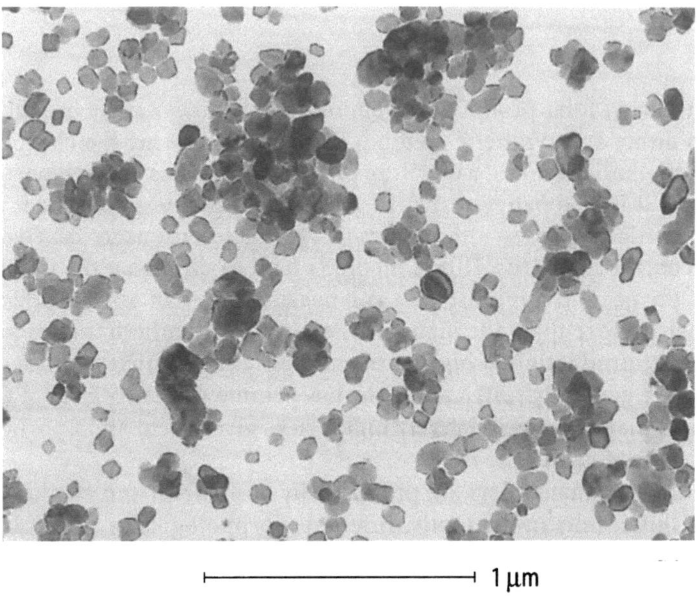

Fig. 2. Transmission electron micrograph of a typical precipitated calcium carbonate filler (courtesy of Zeneca Resins)

3.2 Direct Production Methods

These methods are the principal ones used in the production of fillers based on calcium carbonate and dolomite, clays, talcs, micas and wollastonite. These fillers, on a volume basis, dominate the industry.

In all these processes the starting point is a high grade mineral ore, containing a high proportion of the mineral of interest.This is mined and then treated by a series of purification processes generally described as beneficiation, usually followed by some form of size and/or shape control. These production procedures will vary from mineral to mineral and even from deposit to deposit.

The general steps involved are:

- purification (various physical methods including hand sorting, washing, density separation, flotation, size and shape separation are routinely used, chemical methods such as bleaching are also sometimes employed);
- size reduction and control of particle size distribution (grinding, classification);
- coating (when used);
- dewatering (when aqueous processing has been used);
- drying;
- bagging and labelling.

Further details on some of these processes can be found in the work by Wills [45] and in the section on individual filler types.

3.3 Synthetic Production Methods

Of the various synthetic processes that are available, two are of most relevance in the present context – precipitation from aqueous solution and melt forming. These methods are used where it is not possible to produce adequate products directly from natural sources. This will be because there is no suitable mineral, due to the chemical nature of the product, of particle size and shape requirements, or to purity considerations. The other principal synthetic method in use for filler production is pyrolysis/combustion. This type of process in which the particles are formed in the gas phase is used where very small particles are required, such as with carbon blacks and some silicas. This type of filler is not widely used in thermoplastics and so these processes are not discussed in any detail here, although some information specific to the production of antimony oxide will be found later.

Precipitation from aqueous solution is able to produce high purity materials of a wide chemical range and with very small particle sizes. Because of the formation of the filler as a suspension in an aqueous medium, precipitation methods lend themselves to the incorporation of wet coating procedures. The particles produced by precipitation are generally very small, and difficult to filter and often form a filter cake that shrinks into a hard, difficult to disperse mass on drying. Some coatings, notably fatty acids and surfactants, can reduce these problems [46].

Precipitation and crystal growth are complex subjects with a vast literature. Many factors are important in determining the size and shape of the final particles. These include the reagent concentrations, the method of reagent mixing, the temperature of reaction and of subsequent ageing, the presence of seed particles and trace impurities and the use of habit modifiers. A very useful review has recently been published by Sohnel and Garside [47]. The main applications in the present context are in the production of aluminium and magnesium hydroxides and precipitated calcium carbonates, although the technique is also used to produce a number of other fillers, notably silicas, silicates and barium sulphate.

Aluminium and magnesium hydroxides are difficult to produce directly in any useable form from their natural ores. Filler grade calcium carbonate is widely produced from natural sources, but grinding costs appear to become prohibitive when ultra-fine particles are required and precipitation procedures then become competitive. Further details of precipitation procedures will be found under the specific filler types.

Melt forming is applied to the production of relatively massive particles from silicate glasses, themselves readily produced free from deleterious impurities using cheap mineral sources. The principal attraction of the method is the ability to control particle shape and the principal applications in the present context are in the production of fibres and spheres.

3.4 Filtration and Drying

When wet processing procedures are employed the final product usually has to be isolated in dry powder form before use. This is normally achieved by a combination of filtration and drying processes. In some instances a washing stage also has to be incorporated to remove soluble material, either initially present in the raw material or introduced by the production process (as in many precipitations). All of these procedures can have a significant impact on the process costs and need to be carefully optimised.

One of the aims of the product isolation procedures should be to arrive at a final product form that is easy to transport and handle. Product compaction often occurs during drying, especially when fine precipitated particles are involved. This is generally regarded as a nuisance, requiring some form of comminution in order to obtain a useful product which disperses readily in polymer applications. This compaction process can, in some circumstances, be put to good use and employed to obtain a high bulk density, non-dusty, free flowing powder. This is achieved by various agglomeration procedures, including spray drying, and often incorporating some form of binder.

These agglomeration procedures are still more of an art than an exact science and must strike a fine balance in producing an agglomerated product that is strong enough to survive packing, transport and handling, but is weak enough to disperse readily to the required effective particle size in the polymer application. When binders are used they must be carefully selected to make sure that they are compatible with the polymer system and end application. Most information on agglomeration can be found in the carbon black literature, where pelletisation procedures have been highly developed [48].

3.5 Methods of Surface Modification

3.5.1 *General*

Surface modifiers are used for a number of reasons which have recently been detailed [46]. These include improved production, reduction in moisture content, chemical protection, reduced dustiness and increased bulk density, as well as the more obvious ones of improved composite processing and enhanced composite properties.

The principal types of surface modifier applied during production are fatty acids and organo-silane coupling agents. The methods of application used vary considerably depending on the nature of the surface modifier and the filler production process being used and are generally regarded as proprietary by filler producers.

The different application methods can be broadly divided into dry and wet processes. In addition many coatings can be applied during composite production, in effect using the compounding procedure to carry out the coating. This method is widely used but is outside the scope of this article. The advantages and limitations of the various coating methods have been discussed elsewhere [49].

The aim of the coating process is normally to react the filler surface chemically with the coating agent, although some of the most recent treatments in the patent literature seem to rely on mere physical adsorption, possibly just using the filler as a convenient carrier for what may really be a polymer modifying additive [50]. When fatty acids are being used with some fillers it is also possible for a bulk reaction to occur instead of the desired surface one, leading to the formation of discrete fatty acid salts [51]. This is illustrated in Fig. 3 and applies to both dry and wet coating processes, the conditions of which have to be carefully controlled to minimise this effect.

3.5.2 *Surface Coverage*

All the pre-coating methods in use rely on the addition of a fixed amount of surface modifier to a given amount of filler, thus determining the final composition. It is thus necessary to know in advance how much coating to use for a given application.

The amount that is theoretically needed to completely cover the surface of a filler can be calculated in many instances if the specific surface area of the filler (or better the spacing of coating reactive groups on the surface) and the area oc-

Fig. 3 a,b. Coating of filler particles: a desired uniform surface layer; **b** incomplete layer due to undesirable formation of a separate phase (either the neat coating agent or a product of reaction with the filler material)

cupied by the reacting group on the coating molecule and the coating orientation is known. Thus it is well documented that the area occupied by a stearate molecule oriented perpendicular to a surface is approximately 21 Å^2 [52]. On this basis, a vertically oriented, dense packed layer would need about 0.25 wt% of this coating for each square mg^{-1} of available specific surface. This footprint area has been determined for a number of fatty acids [52] and for a few of the organo-silane surface modifiers [53] and this information can be used to perform similar calculations for them.

While this sort of calculation gives a useful guide to the level of coating likely to be used, the final determination is usually carried out empirically, adjusting the amount of coating until optimum performance is obtained. There are many reasons for this:

a) the desired effect may be optimal at levels of less or more than a complete monolayer, an example being fatty acid coatings on precipitated calcium carbonate, where commercial products seem to have about half of a theoretical monolayer;
b) many of the coatings are expensive and only the minimum amount necessary to achieve the required effect is used, even though the absolute effect may increase with more coating;
c) not all of the specific surface area as determined by nitrogen adsorption may be available for reaction – this can be because some is in pores too small to be accessed by the coating molecules which are considerably larger than the nitrogen molecule that is usually used to determine the surface area, or because of a shortage of reactive sites [54];
d) the coating may exhibit a different orientation from that assumed in making the calculation of footprint area – thus fatty acids can adsorb both flat as well as perpendicular to the surface [55], while organo-silanes can exhibit complex surface structures [56]; some of these concepts are illustrated in Fig. 4.

A number of simple laboratory tests can be used to ascertain the level of coating that may be optimal in practice [57].

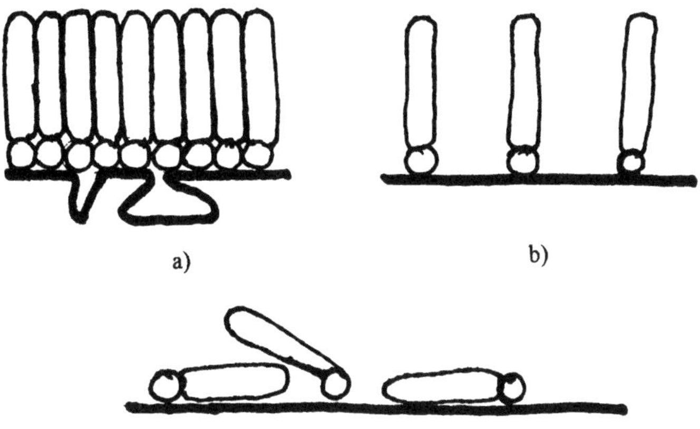

a) b)

c)

Fig. 4 a–c. Some ways in which the actual and theoretical coating monolayer amounts may differ: **a** some of the surface not available to the coating agent; **b** wide spacing of surface reactive sites; **c** orientation of the coating molecule is different from that assumed. (The coating agent is depicted as having a reactive head and an unreactive tail)

3.5.3
Use of Fatty Acids

Fatty acids can be applied by both wet and dry coating methods. A whole range of fatty acid treatments can be envisaged as being used. In practice cost/performance considerations mean that only a relatively few types are encountered. Again for cost purposes, fatty acid blends are used rather than pure ingredients. The most common blend is probably one described as stearine or stearate, being centred around the stearic acid C18 composition. The actual composition of these blends can vary considerably and the stearic acid content can be below 50%. The composition of a typical fatty acid blend such as might be used for filler treatment is given in Table 2.

Table 2. Composition of some typical fatty acid blends of the type used for filler treatment

Fatty acid carbon number	percentage in the blend		
	A	B	C
C12–14 saturated	2	2	2
C15 saturated	–	1	1
C16 saturated	10	29	29
C17 saturated	1	2	2
C18 saturated	34	66	65
C18 unsaturated	–	–	1
C20 saturated	30	–	–
C22/24 saturated	25	–	–

Again, the actual composition used is regarded as proprietary by most filler producers, as the exact nature of the blend can be important in some applications. The colour of the fatty acid blend is one important factor, as this can carry through into the final composite. The level of unsaturation present is one factor that influences the colour. Unsaturation can also have an impact on other application properties, such as polymer ageing and is often an important property to be minimised in selecting a fatty acid blend. In some applications high levels of unsaturation seem to be preferred and fatty acid blends based around the oleic acid composition are then used. Such coatings are reported as being particularly effective for magnesium hydroxides when used in polypropylene [9].

3.5.4
Use of Organo-Silanes

Organo-silane surface modifiers are considerably more expensive than fatty acids but, unlike them, they can considerably enhance filler to polymer bonding and hence result in cost effective increases in performance in many applications. Organo-silane surface modification is a complex subject, but several good reviews are available [58–60]. Organo-silanes are multifunctional materials containing groups capable of forming strong bonds with both the filler surface and the polymer matrix (coupling agent effect). A variety of commercial products exist with reactive functionality tailored to suit different polymer types as illustrated in Table 3. The filler reactive group in these additives is an alkoxy silane. As a result these materials can only react to any degree with fillers containing surface hydroxyls. Thus they are effective on glass, metal hydroxides, clays and other silicates but not on carbonates [61].

While organo-silane treatments are extensively used in both thermoset and elastomer applications, their use in thermoplastics has so far been somewhat restricted. This is because they do not react with the surface of calcium carbonate, one of the principal fillers used in this type of polymer and because of the lack of a suitable reactive functionality for most of the thermoplastic polymers. Today they are principally used in conjunction with glass fibres, calcined clays, aluminium and magnesium hydroxides, micas and wollastonite. The main thermo-

Table 3. Reactive functionalities carried by the principal organo-silane coupling agents for use in thermoplastics

Type of reactive functionality	Principal polymer types for use in
Vinyl	Polyolefins
Methacryl	Polyolefins and styrenics
Amino	Polyolefins, PVC and polyamides
Cationic styrylamine	Most
Chloroalkyl	Polyamides and styrenics
Azido	Polyolefins

plastic polymer type where they are effective is in polyamides, where amino functionality is able to provide a strong polymer bond.

So far there has been little use in the volume thermoplastics such as PVC and polyolefins. The main exception would seem to be the use of amino-silane treated glass in conjunction with maleated polypropylene in polypropylene compounds.

Complex organo-silane products are now emerging which seem able give significant improvements in the impact strength of highly filled polyolefins, especially when used in conjunction with metal hydroxide flame retardants [62]. This may significantly increase their use.

As with fatty acids, organo-silanes can be applied by both wet and dry coating methods. In the present context, wet coating is principally restricted to glass fibre coating, where it can be incorporated into the size, which is already applied from water. Dry coating is the preferred method for coating mica, wollastonite clays and metal hydroxides.

3.5.5
Wet Coating Methods

Wet methods fit naturally into production processes in which the filler is handled in aqueous suspension and the additive is water soluble or water dispersible. In some such cases no extra processing costs are incurred and the coating may even reduce costs by improving filtration and drying characteristics [46]. Such wet coating procedures are widely used with saturated and unsaturated fatty acids which can be readily solubilised as the ammonium or sodium salt. In the present context they are suitable for use with calcium carbonates (especially precipitated ones) and aluminium and magnesium hydroxides. The wet application of silanes seems to be restricted to the special case of glass fibres where, as mentioned above, they can be applied alongside the essential sizing components that are used to protect the fibre during handling [63].

The solution chemistry of fatty acid salts can be quite complex, due to their tendency to form micelles. The solution coating process itself is also complex, with many factors controlling the final coating distribution and structure. Unfortunately little has been published on this topic, which filler producers regard as valuable proprietary information. Variables which can be imagined to have an important effect include: the exact nature of the fatty acid used, the nature of the salt (sodium or ammonium salts are usually preferred), the concentration of the filler and fatty acid salt, the temperature and pH at which the coating is carried out and the conditions of mixing. The sodium salts have the advantages of being more soluble and more pleasant to handle than the ammonium ones. Unless extensive washing is used (an expensive procedure) they do leave traces of sodium in the product however. This can be a problem in some applications, particularly where water adsorption is to be minimised. When the ammonium salt is used, the ammonia is lost during drying and does not contaminate the product.

Effluent disposal is a particular feature which has to be taken into account with wet coating processes. It is important to arrange the coating process such that as much as possible of the coating agent is adsorbed onto the filler, as the remainder is carried through in the filtrate and may make it difficult to dispose of. This can be a particular problem with commercial blends of fatty acids where some of the trace components might not be filler reactive. Much of the solubilising ion (sodium or ammonium) will also be found in the filtrate and this can rule out the use of ammonium salts in some situations.

3.5.6
Dry Coating Methods

Dry coating is probably the main method used in preparing products for thermoplastic applications. Due to the cost of additional drying operations, it is the method of choice when the filler itself is produced by a dry process. It is also useful where wet coating procedures would give rise to effluent problems.

Dry coating is extensively used with fatty acid treatment of natural calcium carbonates. The challenge is to convert as much as possible of the coating to a bound surface layer, with as little unbound salt and remaining free acid as possible. There is little scientific literature on this procedure but some useful studies have been made[51, 64]. A number of different methods are employed. In most cases, unless a small amount of solvent is used, it is necessary for the procedure to be carried out at a temperature where the fatty acid blend is molten. With stearate mixtures this is about 80 °C. Some fatty acids such as iso-stearic acid have the advantage of being liquid at room temperature, but are not widely used as they are more expensive.

In some instances the filler grinding and coating processes can be combined into a single step. This is carried out in ball or attrition mills, where the milling process alone can often provide sufficient heating. The residence time must be sufficient for the reaction to go to completion and the amount of coating added must be carefully controlled. The process can give a very uniform surface layer and there is some indication that the coating may aid the comminution process.

Treatment of pre-ground fillers is also common, and is the method most often used in laboratory studies. This is carried out in a high speed blender such as a Henschel, where the intensive mixing can quickly raise the temperature to that needed for efficient coating. As with the previous process, sufficient time must be allowed for the reaction to go to completion and the amount of coating added must be carefully controlled.

4 Filler Characterisation

4.1 General

Many of the chemical and physical properties of mineral fillers are important in their application in thermoplastics. These include purity, specific gravity, hardness, electrical, thermal and optical properties, surface area, particle shape and size. The determination and importance of many of these has been covered in several reviews [65, 66]. Only a brief coverage is given here for the less ambiguous properties such as specific gravity, hardness and standard thermal and optical properties, with most attention being concentrated on properties such as size and shape which have been found to give particular problems in measurement and interpretation.

4.2 Purity

While always desirable, high purity in itself is not always essential or achieved in many applications. Many impurities have little or no detrimental effect, even when present at high levels, while others can be deleterious at the ppm level. Impurities can be deleterious for a number of reasons including:

a) an adverse effect on filler and composite colour – an example is the presence of organic residues in aluminium hydroxide;
b) increased abrasiveness – quartz is a frequent cause of this sort of problem;
c) increased water adsorption and reduction in electrical properties – this is often the result of the presence of traces of soluble salts;
d) reduction in the heat and light stability of the polymer matrix – traces of certain transition elements, notably copper, can cause this sort of problem;
e) health hazards – examples are the concern over the possible presence of asbestos traces and more recently crystalline silica, in fillers of mineral origin.

4.3 Specific Gravity

Thermoplastics have specific gravities in the range 0.9–1.4. Most mineral fillers have considerably higher specific gravities in the range 2.3–2.8 and their incorporation increase the specific gravity of the composite quite considerably. Except in special cases this is generally of no benefit and is often a disadvantage.

4.4 Hardness

The hardness of mineral fillers is of considerable importance in thermoplastics applications. In general soft fillers are preferred, as the harder ones tend to cause extensive wear of compounding machinery and can lead to degradation of fibre length when used in conjunction with glass fibre reinforcements.

Hardness in the present context is generally measured according to the Mohs scale, developed for classifying minerals and based on the ability of one mineral to scratch another. The scale goes from talc (softest) with a hardness of 1, to diamond with a hardness of 10. In using this scale it must be remembered that it is approximately logarithmic, not linear [67].

4.5 Thermal Properties

The principal thermal properties of importance in the present context are specific heat, thermal conductivity, coefficient of expansion and stability.

4.5.1 *Specific Heat*

This is mainly important in determining the rate at which composites will heat up and cool down during moulding processes. The specific heat of most mineral fillers is about half that of thermoplastics, but when this is converted to the more appropriate volume basis, it is found that there is little difference between fillers and polymers.

4.5.2 *Thermal Conductivity*

This is also important in determining processing behaviour. The thermal conductivity of most mineral fillers is about one order of magnitude higher than that of thermoplastics and their incorporation considerably increases the conductivity of a composite. This effect is beneficial in processing as mouldings can be expected to heat up and cool down more rapidly, leading to shorter cycle times [68].

4.5.3 *Coefficient of Thermal Expansion*

The coefficents of thermal expansion of mineral fillers are considerably less than those of thermoplastic polymers and thus their incorporation can significantly reduce that of a composite material. This is a generally useful effect. High aspect ratio fillers, when aligned by processing, will often give rise to anisotropic effects, leading to problems of warpage [69].

4.5.4
Thermal Stability

It is self evident that mineral fillers need to be stable at the temperatures (up to 350 °C) experienced in processing thermoplastics. Most fillers are stable to much higher temperatures and so this is not usually an issue. However, it is a very important topic for flame retardant fillers which function by decomposing endothermically with the release of inert gasses. To be effective, this decomposition must occur near to the temperature at which the polymer begins to decompose and release flammable volatiles. This is usually not too much above the processing temperature in the case of thermoplastics and hence the exact temperature at which decomposition commences is of great importance. The size and position of the endotherm and the rate at which the inert gas is released are also of importance to the flame retardant effect itself [23].

The determination of thermal stability in this context is a complex topic and has been discussed in depth [23]. Isothermal methods are preferred for detecting the onset of decomposition, as these are more sensitive and approximate more closely to the conditions experienced during processing. Differential scanning calorimetry and thermogravimetric analysis (optimally with evolved gas analysis) are most useful for characterising the decomposition itself [70]. Most of the decomposition reactions that occur with flame retardant fillers are retarded by the application of pressure, such as can occur when they are encapsulated into a polymeric matrix or in the centre of large filler particles and this can even change the decomposition pathway [71]. It has also been reported that the presence of surface treatments can significantly affect the thermal stability [25] as can the presence of various impurities [72–74]. These effects must be taken into account in determining the stability and potential usefulness of various fillers. While these measurements are useful in explaining effects, they are of limited use in predicting the results of flame retardant tests.

4.6
Optical Properties

The refractive index is the most important optical property and its effect in determining the appearance of the polymer composite has already been referred to above. Amorphous fillers such as glass fibres and beads have only one refractive index, but most mineral fillers are crystalline and have anisotropic crystal structures resulting in a number of different indices, and this can cause complex and undesirable interference effects [27].

4.7
Morphology (Particle Size and Shape)

4.7.1
General

Morphology is one of the key factors determining the performance of mineral fillers in all polymers including thermoplastics. While an apparently simple concept, it is very complex in practice, and probably accounts for a great deal of the problems and misconceptions that are experienced in this field. A detailed discussion of the subject has recently been given by Rothon and Hancock [75]. Because of its importance in determining the production methods that are used and the choice of filler types its characterisation is covered in some detail here.

Two points have to be stressed before considering the measurement of morphology. The first point to make in discussing filler morphology is that, except for rare instances such as monomodal glass spheres, the morphology of filler particles is complex and they will have a distribution of shapes and sizes which cannot be expressed as a single parameter.

The second point to make is the importance of defining the effective particle in any given application. The effective particle concept is a very important one. Due to their small particle size and methods of manufacture, the ultimate particles (even where such can be defined) in the dry powder form of many mineral fillers exist in both strongly and weakly bound structures. These are commonly referred to as aggregates and agglomerates respectively. (Note that in the present work the term aggregates will be applied to strongly bonded particle clusters and agglomerates to weakly bound ones. This appears to be the main terminology in use, although the reverse is not infrequently found and appears to be a more accurate derivation from the meaning of the two words [76]). Measurement methods tend to determine either the ultimate particles or the strongly bound aggregates. In practice the aggregates often fail to break down fully during polymer compounding and the partially broken down forms become effective particles in the present sense [42, 75]. An idealised picture of the various types of particle that might arise is illustrated in Fig. 5.

Polymer compounding will cause its own degree of agglomerate and aggregate breakdown, which in turn can vary from system to system, and the aim of morphology determination should be to use measuring conditions that give a similar answer. This is not usually attempted and caution should be used with published values, especially for the finer synthetic fillers, where the problem tends to be most acute.

In principle, the effect of increasing severity of dispersion method on the measured particle size distribution ought to provide useful information on the degree to which a given filler will disperse under various processing conditions. Some work has been carried out along these lines with promising results [77]. The use of this approach, together with measurement techniques that examine different states of particle subdivision, can allow a good picture of the types and

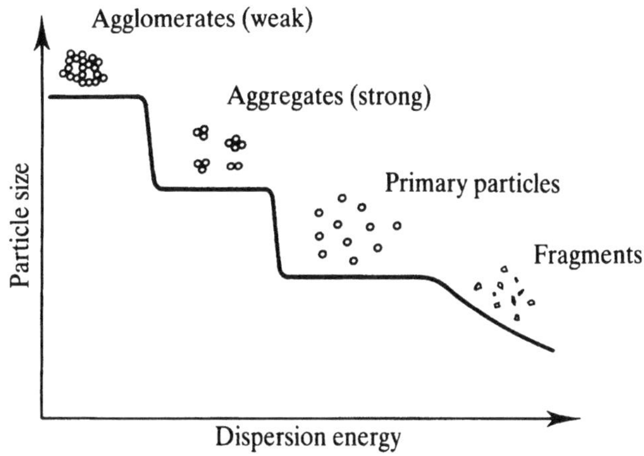

Fig. 5. Idealised view of the way that filler particles disperse and of the different particle types that might be encountered. (Reproduced from [3] with the permission of Addison Wesley Longman Ltd)

sizes of effective particles likely to be found in a given filler type to be determined, as shown in Table 4.

In general, the filler industry recognises these limitations, and tries to use a few relatively simple parameters that, taken in combination, give an approximate, working definition of morphology appropriate to the application in mind. The parameters that are most likely to be encountered are specific surface area, average particle size, effective top size and oil adsorption. The measurement and application of these are discussed in more detail below.

Table 4. Effect of measuring conditions on the apparent particle size of a coated precipitated calcium carbonate (Reprinted from [3] with permission of Addison Wesley Longman Ltd)

Measuring method	Average particle size (micron)	Comments
Laser diffraction of organic dispersion using weak ultrasonics	20	Detecting agglomerates
Laser diffraction of organic dispersion using medium ultrasonics	4	Detecting basic aggregates
Laser diffraction of organic dispersion using strong ultrasonics	0.2	Detecting some form of subaggregate structure
Electron microscopy	0.07	Detecting primary crystallite size
X-ray line broadening	0.07	Detecting primary crystallite size

4.7.2
Particle Size

The most widely used parameter for fillers encountered in the thermoplastics area is particle size. Most fillers are anisotropic in shape and this causes difficulties in defining a simple size parameter. The approach normally taken with sizing methods is to convert to what is known as the equivalent spherical diameter (esd). This is the diameter of a sphere having the same volume as that of the particle. The output of most particle sizing methods is automatically in this form, the major exception being sieves, the results from which need to be modified to arrive at an esd value. The information given in data sheets and many technical publications is often just an average value, which for most fillers used in thermoplastics is generally in the range 0.1–10 microns. Sometimes some idea of the size distribution is also given, generally by providing a top and bottom size value.

Average size is a readily grasped concept, being the size below which 50% of the particles will be found, and may be obtained by a variety of different methods, including sieving, light scattering, sedimentation, microscopy, particle counting and even X-ray line broadening. The methods frequently do not agree well with each other and each filler type seems to have its own preferred method, depending on where it falls in the size range and the needs of the end using technology.

In most instances the average is defined on a weight basis, but in some instances a number basis is employed and this can cause a considerable change in the average value. The most meaningful method of presentation will depend on the end application.

With few exceptions, mineral fillers do not have a unique particle size, but exhibit a range of sizes, the shape of the distribution being determined by the nature of the filler and the method of production. It must be recognised that fillers with the same average particle size can have widely differing distributions and it is the distribution that is generally more important in determining the effects in polymer composites. As an example of the effect of particle size distribution, the strength of thermoplastics seems to be critically affected by the presence of small proportions of material much above 10 microns [36]. Most measuring methods obtain the average value from a determination of the size distribution and where possible this distribution should be obtained and used for serious scientific investigations. Nevertheless, the average size is of considerable use when one is dealing with small changes within one filler type.

Very small amounts of certain sized particles (generally, but not always, on the high side) can be particularly important in some applications. The presence of these is not always detected by the standard sizing methods or apparent from a particle size distribution, even where a top and bottom size is given. The presence of such particles is often measured and quoted separately using a technique most suited to the application in hand. Phrases such as “less than 0.05% w/w above 50 micron” are frequently met with in specifications.

Specific surface area is also frequently used as a defining parameter for the filler types encountered in thermoplastics applications. This is the amount of surface possessed by one gram of the filler material, and in the present context usually lies in the range 1–10 $m^2 g^{-1}$. In some cases this information is calculated from the particle size information, assuming a particular particle shape (usually spherical), but it is today more usually measured by techniques such as gas adsorption [78] or dye adsorption [79]. In the adsorption methods an adsorbing molecule of known covering power is deposited in a controlled way such that a monolayer is present on the filler surface and from the amount adsorbed the specific surface area of the filler can be calculated. The most rigorous method is the BET procedure using nitrogen as the adsorbing molecule. This is a very reproducible procedure, but requires relatively expensive equipment not always possessed by filler producers. The value determined by the BET method can exceed that from dye adsorption as the nitrogen molecule is considerably smaller than a dye molecule and hence is able to access surface that is in small pores and not determined by the dye. In using data measured on a filler powder, it must be remembered that particle breakdown, generating additional surface, may occur during compounding.

The specific surface area is obviously related to the particle size distribution of the filler and can be used as a guide to this. For materials of the same density and shape a higher specific surface area means a smaller particle size, but again it must be remembered that two distinctly different particle size distributions can give rise to the same value for the specific surface area and so it is not a unique property.

4.7.3 *Particle Shape*

As mentioned above, particle shape is an important factor determining the use of fillers in thermoplastics. Anisotropy, or aspect ratio, is particularly important, being valuable in improving factors such as stiffness and heat distortion temperature. Despite this, shape itself is generally only roughly described in filler literature and specifications. This is due to the difficulty of carrying out meaningful measurements quickly and at reasonable cost. Short glass fibres are an exception, where, because of their nature, microscopy coupled with image analysis is able to give a reasonable measure. Again, considerable fibre breakdown can occur during processing and the effective size should be determined in the composite. One relatively simple approach that is sometimes effective is to compare the specific surface area as determined by nitrogen with that calculated from the particle size distribution assuming spherical particles. If these two values agree then the particles are assumed to have a low aspect ratio, whereas if the measured value is greater than the predicted one, the filler is either porous or has some degree of anisotropy. This concept is illustrated in Fig. 6. It is also claimed that some light scattering methods currently used for particle size measurement can be adapted to give shape information directly [80]. The use of fractal analy-

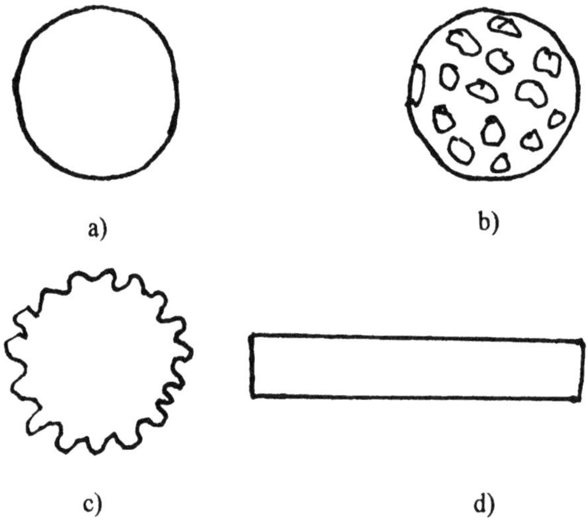

Fig. 6 a-d. Schematic illustration of the ways in which the BET specific surface area can differ from that derived from particle size measurements: All particles have a similar equivalent spherical diameter but in: **a** the particle is solid; **b** the particle is porous; **c** the particle has a rough surface; **d** the particle is anisotropic

sis is now beginning to be explored by the filler industry, especially for carbon blacks [81] and may ultimately allow particle shape to be measured and specified in a useful way.

One important consequence of the particle size and shape distributions is the ability of the filler to pack and fill space. This ability determines the effective filler volume fraction that can be incorporated before filler-filler interactions become dominant. A high packing ability is needed when high filler loadings are required without too much loss in polymer properties, while a low packing ability is beneficial when effects such as electrical and thermal conductivity that rely on a network structure (percolation) being established are desired. An excellent review of the effects of particle size and shape distributions on packing has been published by German [82].

Oil absorption is a very simple technique which when carefully applied can give a useful guide to the packing ability of fillers [83]. This determines the amount of a selected oil that is needed to just form a continuous phase between the filler particles when they are subjected to a certain mixing procedure. This is a good guide to the maximum packing fraction of filler that is likely to be achievable in a polymer matrix, especially if the oil used is chosen to have a similar polarity to that of the polymer to be used.

5 Production, Properties and Applications of the Principal Filler Types

5.1 Natural Products

5.1.1 *Calcium Carbonate and Dolomite*

Ground natural calcium carbonate is, on a volume basis, the main filler material used in thermoplastics. The principal applications by far are in PVC, but significant quantities are also used in other polymers, notably polypropylene and polyamides. A wide range of particle sizes is used according to the nature of the application.

The principal reasons for its use are relative abundance, low cost of producing relatively fine, low aspect ratio particles of high purity and the ready improvement in properties obtained on surface modification with low cost fatty acids.

Calcium carbonate has a number of crystal modifications, but the calcite form is the one that is principally used for filler applications. Pure calcite is a relatively soft material (Moh hardness 3.0) with a specific gravity of 2.7.

Hancock and Rothon have given a good recent account of the origin and production of natural calcium carbonate fillers [84]. Calcite deposits are commonplace, but differ in their exact origin and it is this which is the principal factor determining their properties and detailed method of production. The principle forms used for filler production are chalk, limestone and marble. Chalk is a fairly soft material formed from the sedimentary skeletal remains of sea animals. These remains consist of low aspect ratio calcite crystals a few microns in size which are sufficiently loosely bound together to be readily separated by milling procedures. Limestone is similar to chalk, but is harder and denser having been compacted by a variety of geological processes. Marble is even harder and denser, being a limestone that has been subject to high pressures and temperatures resulting in extensive recrystallisation.

Processing of all the forms is by simple mining procedures followed by wet or dry grinding and classification. Depending on the purity of the deposit, magnetic and/or flotation methods of beneficiation may be utilised.

A wide range of particle sizes is produced to suit different applications. Initially milling only has to break the relatively weak bonds between crystals of calcite, but for the finer products fracture of these crystals has to occur. Dry milling is used for the coarser fractions, but milling costs rise steeply as the degree of fineness increases until it becomes more economic to use wet milling, despite the additional drying costs. Eventually, as the required particle size is reduced even further, precipitation becomes the most economic procedure.

Much of the natural calcium carbonate used in thermoplastics is fatty acid treated. Manufacturers give little detail about their coating processes but it is likely that both wet and dry coating procedures are utilised.

Dolomite is an alternative mineral that is used in some regions in place of calcite for certain applications. Dolomite is a calcium magnesium carbonate ($CaCO_3.MgCO_3$) and occurs widely in nature. Although generally similar to calcite in properties, it is slightly harder (3.5), denser (2.85) and more acid resistant. Production is similar to that for calcite, but milling is more costly and it tends to only be available at the coarser end of the size spectrum.

5.1.2
Clays

Clay minerals are widely used in the filler industry and the production of the various types has been described by Hancock and Rothon [85]. The products available range from unaltered kaolin to products produced by calcination. Kaolin deposits are widespread throughout the world . While simple clay minerals are extensively used as fillers in elastomers, their use in thermoplastics is more restricted and the principal products used are those obtained by calcination of kaolinite. Thus metakaolinite finds application in PVC and silane treated calcined clay in polyamides. The production of these forms only is discussed below.

A number of different products can be obtained by calcination of china clay (kaolinite, $Al_2O_3.2SiO_2.2H_2O$) and two of these find use in thermoplastic applications.

The first is metakaolin. This is a partially calcined product that forms above about 500 °C. Only about 10% of the original hydroxyl groups of the kaolinite are retained and much of the crystalline nature of the structure is destroyed. Metakaolin is considerably more reactive than the original kaolin and appears to have an especially reactive surface. It is generally used uncoated and finds most use in plasticised PVC cable insulation, where it is reported as giving uniquely useful electrical properties [86].

The second material is simply known as calcined clay and is formed when kaolin is heated above 1000 °C. This is an amorphous material with a defect spinel structure. A few isolated hydroxyls are retained on the particle surface and this enables the material to react with materials such as organo-silanes.

The principal use of calcined clay in thermoplastics is in polyamide moulding compounds, for which purpose it is normally coated with an amino functional organo-silane.

5.1.3
Talcs

Talc is a crystalline form of magnesium silicate which when pure has the formula $Mg_3(Si_4O_{10})(OH)_2$. The uses of talc in polymers are largely determined by its crystal structure. This is a lamellar structure based on magnesium hydroxide sheets sandwiched between two siloxane layers. Bonding between the layers is by weak Van der Waal forces and hence talc is a very soft material that readily undergoes cleavage to form high aspect ratio particles. The Moh hardness is 1, and the specific gravity is 2.8. In thermoplastics the principal use of talcs is in polypropylene where the high aspect ratio gives significant improvements in stiffness. Despite being the principal factor for its use, the aspect ratio of most commercial talcs is not specified.

There are several good reviews on the production and properties of talc[87, 88]. In practice a number of related minerals are produced and marketed under the name of talc. For polymer applications only two minerals are of significance – talc itself and chlorite (chlorite is a mixed magnesium silicate/hydroxide with some iron and aluminium substitution). Commercial products will often contain mixtures of the two minerals in different proportions. Some related minerals, such as tremolite, are of health concern and care is taken to use deposits that do not contain them. After mining, the crude rock is crushed, beneficiated, milled and classified. Benefication may include hand or mechanical sorting and in some cases froth flotation. The milling is a very important step as it not only determines the particle size, it delaminates the particles and hence determines the aspect ratio. Various milling procedures are in use. Despite much research, surface treatments are of little use in most talc applications and virtually all talc for thermoplastics use is untreated.

5.1.4 *Micas*

Mica is the name given to a group of related silicate minerals having the generalised formula $KM(AlSi_3O_{10})(OH)_2$, where M can be Al, Fe or Mg. As with talc, the use of mica in polymers is a result of its crystal structure. Like talc it is a layered structure and readily cleaved into high aspect ratio sheets. The bonding between the layers is however stronger than with talc and hence the particles are harder and have greater structural integrity. Commercial micas vary somewhat in chemical composition and Moh hardness varies from 2–2.5 and specific gravity from 2.75–2.90. The principal use of mica in thermoplastics is as a reinforcing filler, where it is used to some extent in most types.

The production of mica for polymer applications has been reviewed by Hawley [89]. The aim of the processing is to purify the deposit and to produce particles of relatively small diameter with an aspect ratio of 50–200. The natural minerals are generally of much larger size than required and so the milling has both to delaminate and fracture the particles. The milling is the key process and a variety of methods, both wet and dry, are used, accompanied by various classification methods. Surface modification is important in many mica applications and a variety of treatments are used, especially organo-silanes. The methods of treatment are generally not disclosed.

5.1.5 *Wollastonite*

Wollastonite is a form of calcium silicate ($CaSiO_3$). It is a relatively hard (Moh hardness 4.8) white material with needle-like particles. The specific gravity is

2.9. The main interest in wollastonite in polymer applications is due to its acicular nature which gives it reinforcing properties. Principal uses in thermoplastics are in polyamides and polypropylene. Only a few exploitable deposits exist and this may have inhibited the growth of polymer applications which tend to require the highest aspect ratio material.

The production, properties and applications of wollastonite have been reviewed by Copeland [90]. Production is by mining and crushing of suitable wollastonite bearing rocks followed by magnetic and/or froth flotation . The needle-like particles are then obtained by grinding. The inherent aspect ratio of wollastonite particles is not very high and as much of this as possible must be preserved for polymer applications. For this purpose attrition milling and classification is used. Commercial products for thermoplastic polymer applications generally have the highest aspect ratios (10–20:1). The lower aspect ratio products find other applications, which enhances the economics of the overall operation.

A variety of surface treatments, many proprietary, are offered with wollastonite fillers.

5.1.6
Hydromagnesite/Huntite Mixtures

Hydromagnesite is a basic magnesium carbonate {$4MgCO_3.Mg(OH)_2.4H_2O$}, while huntite is a mixed calcium magnesium carbonate {$3MgCO_3.CaCO_3$}. Both materials generally exhibit a platy crystal habit.

These two minerals always seem to occur together in fairly equal proportions and to date no significant high purity deposit of either mineral alone has been identified. Significant deposits containing the mixtures exist in some areas, notably Greece, and the mixed product has been marketed commercially as a filler for thermoplastics for some time. Filler grade product is made by simple mining, drying, milling and classification procedures. Fatty acid coated grades are available, presumably produced by a dry coating method.

This mixed product consists of small, platy particles with a relatively high surface area (15–20 $m^2 g^{-1}$). The principal interest has to date been as a flame retardant filler, principally for polypropylene. Both component phases decompose endothermically with the release of inert gas at relatively low temperatures. They are stable enough to allow incorporation into polymers such as polypropylene, but not polyamides. The performance of the two phases alone and in combination in polypropylene has been reported [91]. As expected from their thermal properties, hydromagnesite was the more effective flame retardant. The decomposition pathway of hydromagnesite has been shown to be considerably affected by pressure and this may affect its flame retardancy [71].

5.2
Synthetic Products

5.2.1
Glass Fibre

Their are a number of good reviews dealing with the production and use of glass fibres in polymer applications [92–94]. The first commercial use of glass fibres in thermoplastics appears to been in the early 1950s. Today they are one of the principal filler types in use. Although relatively expensive compared to other fillers, the use of glass fibre is attractive because it produces considerable increases in strength and stiffness, including heat distortion temperature, without too much loss in toughness, and the resulting compounds are cost effective for their applications. The main disadvantages of their use are the reduction in toughness mentioned above, anisotropic composite properties, including shrinkage, increased processing costs and enhanced machine wear due to their abrasive nature.

Surface treatments are almost invariably added during the manufacture of glass fibres and play an important role in determining their processing and reinforcing characteristics.

Because of problems with feeding to the melt and preserving fibre length during compounding, the fibre forms in use in thermoplastics are mostly of short length (chopped strand or milled fibre). Despite this they are still able to produce valuable effects.

A wide range of glass compositions can, in principle, be used for fibre production. In practice most glass fibre used for polymer applications is produced from one of the commonest and hence cheapest suitable glasses, known as E glass. This is a lime-alumina-borosilicate glass with a good balance of properties. Other glass compositions are used when enhanced performance in some specific area can justify the extra cost (e.g. higher tensile strength or corrosion resistance).

The glass batch is first prepared by melting the appropriate ingredients in a high temperature furnace. The molten glass is then drawn through bushings containing a large number of holes, or spinnerettes, at high speed (several thousand meters per minute) where it is shaped and stretched, achieving a final diameter of between about 5 and 20 microns depending on the drawing conditions. The spun fibre is then cooled with a water spray and an aqueous based size is applied before the individual filaments are gathered into strands for further processing.

The size is a complex mixture of chemicals, the exact composition of which varies according to the final application for the glass and is treated as proprietary by the manufacturers. It serves a number of important functions. The first is to protect the fibre surface from abrasion and other damage. The virgin fibre is almost flaw free and hence has a very high strength as first made. This virgin surface is readily damaged however, and must be protected if the strength is to be retained. The second is to act as a binder for the individual fibres giving good handling characteristics and controlled dispersion in the polymer matrix. The third is to act as a lubricant during fibre and composite processing. The fourth and final function is to provide good wettability and optimum strength of bonding with the matrix polymer. This final function is usually achieved by the use of organosilane coupling agents, the nature of which have to be varied according to the identity of the host polymer.

The short fibre forms used in thermoplastics are produced by cutting the fibre strands into lengths of about 3–6 mm or by hammer milling which results in fibre lengths of about 1–3 mm. The glass diameters used in thermoplastics are normally about 15 micron.

5.2.2
Glass Beads

The production and use of solid glass beads has been reviewed by Katz [95]. Solid glass beads are principally used in engineering thermoplastics such as polyamides and ABS. Their round shape allows high loadings, and hence stiffness, to be obtained with least increase in viscosity. It is also claimed that the absence of sharp, stress raising edges gives better toughness retention.

A variety of methods are available for the production of glass beads. These generally involve the atomisation of molten glass or the melting of fine glass powder. A variety of surface treatments are used, mainly of the silane type. A wide particle size range is available, but the finer sizes (30 micron and below) are most used in thermoplastics.

5.2.3
Silicas

There are at least five types of synthetic silicas that can be considered for use in polymers. These are generally known as fumed, arc, fused, gel and precipitated. A detailed review of their production and uses has been given by Watson [96]. The types most often encountered in thermoplastics are the gel and precipitated silicas which are frequently used as antiblocking agents in polymer films and as gloss reducing agents in polymer sheets.

Silica gels are porous particles which are produced by adding sulphuric acid to a solution of sodium silicate. Under the right conditions this results in the formation of a solid hydrogel. This hydrogel is washed to remove the co-product sodium sulphate, dried and ground. Different types of silica gel are produced depending on the drying rate.

Precipitated silicas are also produced by the addition of sulphuric acid to a solution of sodium silicate but under different conditions, which result in the formation of aggregates of tiny discrete particles rather than the massive structure of a gel. After precipitation the slurry is filtered, washed, dried and deagglomerated.

5.2.4
Aluminium Hydroxide

Aluminium hydroxide has a Moh hardness of about 3 and a specific gravity of 2.4. It decomposes endothermically with the release of water at about 200 °C and this makes it a very useful flame retardant filler, this being the principal reason for its use in polymers. The decomposition temperature is in fact too low for many thermoplastics applications, but it is widely used in low smoke PVC applications and finds some use in polyolefins. For these applications low aspect ratio particles with a size of about 1 micron and a specific surface area of 4–10 $m^2 g^{-1}$ are preferred. The decomposition pathway can be diverted through the monohydrate by the application of pressure, and this may reduce the flame retardant effect [97]. This effect can be observed with the larger sized particles. Although it is chemically the hydroxide, it has for many years been known as alumina trihydrate and by the acronym ATH.

The production of flame retardant quality aluminium hydroxide has recently been reviewed [98]. Various crystal forms of aluminium hydroxide exist, but that used for polymer applications is Gibbsite. This occurs widely in nature, usually in the rock bauxite, but the natural form is usually not suitable for direct use and synthetic products are nearly always employed. Most aluminium hydroxide is manufactured through the Bayer process used to make alumina for refractory applications.

In the Bayer process, the bauxite is leached with hot sodium hydroxide, thus forming a solution of sodium aluminate. After purification this solution is seeded with crystals of gibbsite and cooled. The process steps are summarised in Eqs. (1) and (2):

$$Al(OH)_3+2NaOH \rightarrow 2NaAlO_2+4H_2O \tag{1}$$

$$2NaAlO_2+4H_2O \rightarrow 2Al(OH)_3+2NaOH \tag{2}$$

Like many filler production processes, there is no overall chemical change, the reactions being used to reduce particle size and to allow a high level of purification by going through a solution phase.

This Bayer process is operated with the needs of the refractory and metal industries in mind to produce large aluminium hydroxide crystals of about 80 micron. For filler applications these are filtered and washed and then ground to produce different sized fractions. While the large scale of the Bayer process plants confer economic advantages, the process has some limitations as far as producing filler grade products is concerned. The grinding operation tends to cause cleavage along certain crystal planes and so the particles produced in this way become quite platy as the particle size is reduced. There is also an economic limit to the fineness of particle that can be produced by milling. Finally although quite a lot of purification takes place some troublesome impurities remain, principally organic residues and sodium.

The organic residues can cause colour problems, while the presence of sodium can give high levels of water adsorption and hence unstable electrical properties. Various modifications to the basic process are used to overcome these difficulties, although they increase the production costs. Various purification procedures can be employed to remove colour and to reduce the level of sodium contamination, while dedicated precipitation processes are used when smaller or less platy particles are needed.

Many surface modifications are used with aluminium hydroxide, which responds to both silane and fatty acid treatments. Special proprietary silane coatings seem to be preferred for polypropylene applications [99]. Despite the production being water based, the preference seems to be for dry coating procedures.

5.2.5 *Magnesium Hydroxide*

Magnesium hydroxide occurs in nature as the mineral brucite. It has a Moh hardness of about 3 and a specific gravity of 2.4. It starts to decompose endothermically with the release of water at about 300 °C and the principal interest in it is as a flame retardant filler for thermoplastics such as polyolefins and polyamides, where the processing temperature is too high for aluminium hydroxide to be utilised effectively. For thermoplastic applications low aspect ratio particles are favoured with a particle size of about 1 micron and a specific surface area in the range 4–10 m2 g^{-1}.

The development of the different methods for the production of flame retardant grade magnesium hydroxide has recently been reviewed [100]. Although not a common mineral, there are some workable deposits of brucite, especially in the US and China and product obtained by milling high purity brucite deposits is being marketed, but has so far made little impact. This is probably because the high levels needed for flame retardancy can only be tolerated if the particle size and shape are carefully controlled and this requires the use of synthetic methods of production.

A number of inexpensive magnesium-containing sources are available on which to base such a process. Until recently one particular synthetic process has dominated the market but now a considerable number are under development or in actual use. This change is probably due to the high cost of production of the original process and the imminent expiry of the patents associated with that process and product form.

Synthetic magnesium hydroxide is not new. It has been produced on a relatively large scale as an intermediate in the manufacture of refractory magnesium oxide for many years. The principal process has been precipitation based on the addition of lime or dolmitic lime to magnesium chloride solution, usually obtained from sea water or from solution mining of a suitable deposit. Despite intense efforts by most of the producers, this route does not seem able to produce a satisfactory product for wide penetration of the thermoplastics market. Magnesium hydroxide naturally precipitates as very small platy particles with a high tendency to agglomerate. These are not suitable for use in polymers and the challenge, that the above process does not seem to have been able to overcome, has been to develop an economic process that enables the production of more suitable, larger, non-agglomerating, particles.

The initial process to achieve success was developed by Miyata of Kyowa in the mid 1970s [101].This process is based on the hydrothermal conversion in an autoclave of the fine agglomerated particles formed by addition of ammonia or lime to a magnesium salt solution. This hydrothermal conversion results in the formation of particles of about 1 micron in size and with a fairly low aspect ratio. This original process appears to be expensive to operate because of the low value of the ammonium or calcium chloride co-product which has to be disposed of and because the reported reaction conditions give a low yield and a relatively slow reaction.

Recent developments to the hydrothermal process include improvements in yield and reaction rate and in overcoming the difficulty associated with the co-product salt. One method of overcoming the co-product problem is to use magnesium nitrate instead of chloride, with the ammonium nitrate being utilised for fertiliser production [102–104]. At least one plant based on this concept is now in commercial production. While a considerable advance on the initial chloride process, the nitrate route does require close integration with a fertiliser process and thus lacks flexibility. An alternative approach being developed is to recycle the ammonium salt co-product (nitrate or chloride) and use it to leach magnesium oxide, a potentially inexpensive raw material [103].

The other main approach is to use the controlled hydration of magnesium oxide to precipitate the hydroxide in the desired form [105]. This has the advantage of having no co-product salt to consider, but obtaining the oxide in the desired form at a reasonable cost is a problem that can only be overcome in special circumstances. Despite this there are at least two significant commercial producers who use this route.

A number of coatings are offered with magnesium hydroxide. These include saturated and unsaturated fatty acids and silanes. The coating methods have not been disclosed, but presumably wet coating is used where practical. A number of publications, especially from Japan, refer to the use of coatings of various types to protect the filler from attack by carbon dioxide [106], but it is not clear whether this is a real problem and whether these are in commercial use.

Some typical magnesium hydroxide particle forms produced by a version of the hydrothermal process are presented in Fig. 7

5.2.6
Basic Magnesium Carbonate

Synthetic basic magnesium carbonate is produced for filler purposes. This material has a composition that approximates to the formula $4MgCO_3 \cdot Mg(OH)_2 4H_2O$ and is essentially a synthetic form of the mineral hydromagnesite described earlier. It is mostly used as a smoke suppressant in PVC and similar polymers. It is reported in the trade literature as starting to decompose at about 200 °C.

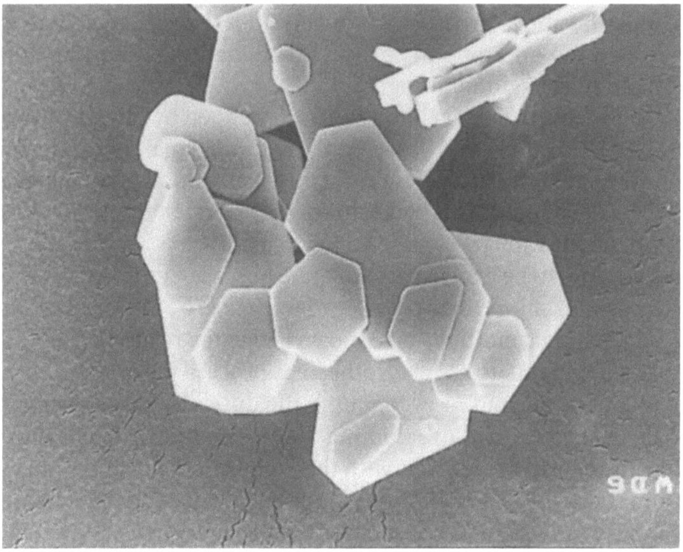

1 μm

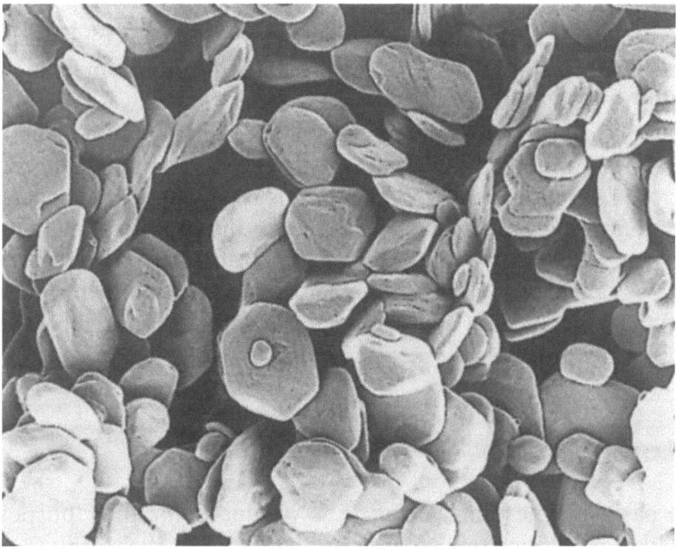

1 μm

Fig. 7. Scanning electron micrographs of two forms of synthetic magnesium hydroxide (Courtesy of Flamemag International GIE)

Synthesis appears to be by carbonation of a crude magnesium hydroxide slurry to form a dilute solution of magnesium bicarbonate. After filtration, particles of basic magnesium carbonate are precipitated by heating this solution. The resulting product has quite a high surface area and oil adsorption and is only suitable for incorporating into polymers at moderate loadings. This probably accounts for its use as a smoke suppressant, where only low loading levels are used, but not as a primary flame retardant where very high levels are required. One must assume that the producers have attempted unsuccessfully to modify the precipitation to produce a lower surface area product for flame retardant applications.

5.2.7
Precipitated Calcium Carbonate (PCC)

In the thermoplastics area, precipitated calcium carbonate is principally used in PVC applications, a market with which it has been associated since the early days of the polymer. Despite some erosion by coated natural products, the combination of small particle size and fatty acid coating continues to give a unique blend of properties in both unplasticised and plasticised PVC formulations. The advantages include easier processing, better surface finish, good low temperature properties and resistance to crease whitening and to scratching.

The production process is able to produce all three crystal modifications of calcium carbonate and a wide variety of particle sizes and shapes, including plates and acicular forms [107]. However, only the calcite form with a rhombic shape and a low aspect ratio seems to have found much commercial application in polymers. For filler applications the particles have an ultimate particle size of 50–100 nanometers, a specific surface area of 15–25 $m^2\,g^{-1}$ and a low aspect ratio.

A number of precipitation processes have been proposed and used in the past including the double decomposition reaction between calcium chloride and sodium carbonate

$$CaCl_2 + Na_2CO_3 \rightarrow CaCO_3 + 2NaCl \quad (3)$$

and the reaction of sodium carbonate with calcium hydroxide (lime – soda process)

$$Ca(OH)_2 + Na_2CO_3 \rightarrow CaCO_3 + 2NaOH \quad (4)$$

The only significant production process today is by carbonation of lime slurry, itself obtained by calcination and hydration of a calcium carbonate mineral. The process is set out below and it can be seen that this is another example of using a chemical method to achieve a particle size reduction:

$$CaCO_3 \rightarrow CaO + CO_2 \quad (5)$$

$$CaO + H_2O \rightarrow Ca(OH)_2 \quad (6)$$

$$Ca(OH)_2 + CO_2 \rightarrow CaCO_3 + H_2O \quad (7)$$

$$\text{Overall } CaCO_3 \text{ (crude)} \rightarrow CaCO_3 \text{ (fine grade)} \quad (a)$$

One great advantage of this process is that there is no co-product salt to remove. This eliminates an expensive washing operation and the need to find a market for, or otherwise dispose of, the co-product. The reaction is also quite exothermic and this heats the product slurry up to a temperature that is convenient for carrying out the coating process. The carbonation process can produce many different forms of PCC, depending on the precise reaction conditions used. Commercial PCC production commenced in the last century and the carbonation reaction is well known as the lime water test for carbon dioxide, but there is little useful scientific literature on the factors controlling particle form. The patent literature gives some idea of the conditions used [108]. Typical calcite crystals made by the carbonation process are presented in Fig. 2

PCC is normally used in fatty acid coated form in thermoplastic applications. The means by which the coating is applied is treated as proprietary by most producers but it is probably by the wet coating method, with a solution of the fatty acid salt added to an aqueous slurry of the filler, followed by filtration and drying. As described earlier, the use of fatty acid coatings can improve product filtration and drying characteristics.

5.2.8 *Antimony Oxides*

Antimony trioxide (Sb_2O_3) and pentoxide (Sb_2O_5) are both used in combination with halogen compounds for flame retarding polymers. Their manufacture properties and use have been described by Touval [109]. The trioxide is the predominate form for use in thermoplastics.

Antimony trioxide occurs in several modifications of which the senarmontite form is the preferred one for polymer uses. This does occur as a natural mineral but the deposits are generally too impure for direct use and virtually all the material used in polymers is made synthetically. The principal synthetic method is to blow air through the molten metal or its sulphide. The metal oxide sublimes from the melt and is recovered by cooling and filtration. Particle size is an important factor governing performance and this is controlled by factors such as the temperature and the cooling rate.

6 References

1. Anon (1993) Plast Rubber Wkly Oct 23:14
2. Hancock M (1995) Filled thermoplastics. In: Rothon RN (ed) Particulate-filled polymer composites. Longman, Harlow, p284
3. Rothon RN (ed) (1995) Particulate-filled polymer composites. Longman, Harlow
4. Jones FR (1994) Handbook of polymer-fibre composites. Longman, Harlow
5. Katz HS, Milewski JV (1987) Handbook of fillers for plastics. Van Nostrand, New York
6. Milewski JV, Katz HS (1987) Handbook of reinforcements for plastics. Van Nostrand, New York
7. Enikopylan NS (1990) Filled polymers 1: advances in polymer science No 96. Springer, Berlin Heidelberg New York

8. Lutz JT Jr (1989) Thermoplastic polymer additives. Marcel Dekker, New York
9. Miyata S, Imahashi T, Anabuk H (1980) J Appl Polym Sci 25:415
10. Hancock M (1995) Filled thermoplastics. In: Rothon RN (ed) Particulate-filled polymer composites. Longman, Harlow, p281
11. Hancock M (1995) Filled thermoplastics. In: Rothon RN (ed) Particulate-filled polymer composites. Longman, Harlow, p284
12. Chow TS (1980) J Mater Sci 15:1873
13. Volenberg PHT, Heikkens D (1986) Polymer 30:1656
14. Ashton JE, Halpin JC (1969) Primers in composite materials. Technomic, Stanford
15. Okuno K, Woodhams RT (1975) Polym Eng Sci 15:4
16. Radosta JA (1976) Soc Plast Eng Technical Papers 22:465
17. Hancock M, Tremayne P, Rosevear DJ (1980) J Polym Sci 18:3211
18. Pukanszky B (1985) 28th Polym Compos Proc Microsymp Macromol 67:2811
19. Janear J (1989) J Mater Sci 24:3947
20. Heikkens D (1986) Filplas '86, BPF /PRI Conference, London
21. Green J (1989) Flame retardants and smoke suppressants. In: Katz HS, Milewski JV (eds) Handbook of fillers for plastics. Van Nostrand, New York, chapter 4, p93
22. Hornsby PR, Watson CL (1986) Plast Rubber Process Appl 6:169
23. Rothon RN (1995) Effects of particulate fillers on flame retardant properties of polymers. In: Rothon RN (ed) Particulate-filled polymer composites. Longman, Harlow, Chap 6, p207
24. Hughes P, Jackson GV, Rothon RN (1993) Makromol Chem Macromol Symp 74:179
25. Ashley RJ, Rothon RN (1991) Plast Rubber Compos Process Appl 15:19
26. Technology of titanium dioxide pigments technical leaflet available from Tioxide UK, Stockton on Tees, Cleveland, TS18 2NQ, England
27. Ferrigino TH (1987) Principles of filler selection and use. In: Katz HS, Milewski JV (eds) Handbook of fillers for plastics. Van Nostrand, New York p25
28. Abe M (1987) Japan Kokai Tokyo Koho, Japanese patent application 62,263,243
29. Goodman H (1992) Filplas'92 BPF Conference, Manchester
30. Kauder OS (1989) Stabilisers. In: Lutz JT Jr (ed) Thermoplastic polymer additives. Marcel Dekker, New York p437
31. Allen NS, Edge M (1992) Fundamentals of polymer degradation and stabilisation. Elsevier, London
32. Bryk MT (1991) Degradation of filled polymers. Ellis Horwood, Chichester
33. Bramuzzo M, Savadori A, Bacci D (1985) Polym Compos 6:1
34. Hancock M (1995) Filled thermoplastics. In: Rothon RN (ed) Particulate-filled polymer composites. Longman, Harlow p287
35. Abolins V (1982) US Patent Application 4,317,761
36. Svehlova V, Polovcek E (1990) Angew Makromol Chem 174:205
37. Rothon RN, Hancock M (1995) Selection and use of particulate materials. In: Rothon RN (ed) Particulate-filled polymer composites. Longman, Harlow, p35
38. McGenity P, Paynter CD, Adams JM (1989) Filplas '89, BPF/PRI Conference, Manchester
39. Woodhams RT (1984) Polym Eng Sci 24:169
40. Hancock M, Pickering FG (1984) CDC Conf London
41. Liauw CM, Lees GC, Hurst SJ, Rothon RN, Dobson DC (1995) Plast Rubber Compos Process Appl 24:249
42. Titow WV, Lanham BJ (1975) Reinforced plastics. Applied Science Publishers, London
43. Suetsegu Y (1990) Int. Polym Process 5(3):185
44. Hutley TJ, Darlington MW (1985) Polym Commun 26:264
45. Wills BA (1992) Mineral processing technology, 5th edn. Pergamon, Oxford
46. Rothon RN (1996) Surface modification of mineral fillers. Loughborough Fillers Symposium, Loughborough, UK
47. Sohnel O, Garside J (1992) Precipitation. Butterworth, Oxford

48. Donnet JB, Voegt A (1976) Carbon black. Marcel Dekker, New York, p8
49. Rothon RN (1995) Surface modification and surface modifiers. In: Rothon RN (ed) Particulate-filled polymer composites. Longman, Harlow, p125
50. Metzemacher HD, Seeling R (1991) Canadian Patent Application No. 2 028 969
51. Gilbert M, Sutherland I, Guest A (1992) Filplas '92 BPF/PRI Conference Manchester, UK
52. Ogino K (1990) Yukugaku 39:398
53. Miller D, Ishida H (1984) Surf Sci 148:601
54. Ottewill RH, Tiffany JM (1967) J Oil Colour Chem Assoc 50:844
55. Kipling JJ, Wright EHM (1964) J Chem Soc 3535
56. Ishida H (1984) Polym Compos 5(2):101
57. Liauw CM, Lees GC, Hurst SJ, Rothon RN, Dobson DC (1995) Plast Rubber Compos Process Appl 24:211
58. Rothon RN (1995) Surface modification and surface modifiers. In: Rothon RN (ed) Particulate-filled polymer composites. Longman, Harlow, p139
59. Ishida H (1985) Polym Sci Tech 27:25
60. Plueddemann EP (1982) Silane coupling agents. Plenum, New York
61. Vanderbilt BM, Jaruzelski JJ (1962) Ind Eng Chem Prod Res Dev 1:188
62. Godlewski RE, Heggs RP (1989) Coupling agents. In: Lutz Jt Jr (ed) Thermoplastic polymer additives. Marcel Dekker, New York, p51
63. Jones FR (1994) Glass fibres-surface treatments. In: Jones FR (ed) Handbook of polymer-fibre composites. Longman, Harlow, p42
64. Fekete E, Pukanszky B, Toth A, Beroth A (1990) J Colloid Interface Sci 135(1):200
65. Rothon RN, Hancock M (1995) General principles guiding selection and use of particulate materials. In: Rothon RN (ed) Particulate-filled polymer composites. Longman, Harlow, p1
66. Katz HS, Milewski JV (1987) Introduction and guide for the selection and use of fillers. In: Katz HS, Milewski JV (ed) Handbook of fillers for plastics. Van Nostrand, New York, p1
67. Ferrigino TH (1987) Principles of filler selection and use. In: Katz HS, Milewski JV (ed) Handbook of fillers for plastics. Van Nostrand, New York, p28
68. Darlington MH, Harrison P, Sanders D (1981) PED Symposium, Loughborough
69. Harrison P, Sheppard RF (1970) Plastics International Conference, Reinforced Thermoplastics. Solihull, UK, October 1970
70. Hornsby PR, Wang J, Rothon RN, Jackson G, Wilkinson G, Cossik K (1996) Polym Degrad Stab 51:235
71. Hancock M, Rothon RN (1995) Principle types of particulate fillers. In: Rothon RN (ed) Particulate-filled polymer composites. Longman, Harlow, p83
72. Wojcik J (1969) Przem Chem 48:11
73. Miyata S (1992) European Patent Application 0,498,566 A1
74. Miyata S (1996) US Patent 5,480,929
75. Rothon RN, Hancock M (1995) General principles guiding selection and use of particulate materials. In: Rothon RN (ed) Particulate-filled polymer composites. Longman, Harlow, p12
76. Kaye BH (1989) A random walk through fractal dimensions. VCH, New York
77. Thoma SG, Ciftiogli MC, Smith DM (1991) Powder Technol 68:53
78. Brunauer S, Emmett PH, Teller E (1938) J Am Chem Soc 60:309
79. Mullin JW, Murphy JD, Sohnel O, Spoors G (1989) Ind Eng Chem Res 28:1725
80. Baudet G, Bizi M, Rona JP (1993) Part Sci Technol 11: 73
81. Herd CR, McDonald GC, Hess WM (1992) Rubber Chem Technol 65:107
82. German RM (1989) Particle packing. Metal Powder Industries Federation, New York
83. Huisman HF (1984) J Coat Technol 56:65
84. Hancock M, Rothon RN (1995) Principle types of particulate fillers. In: Rothon RN (ed) Particulate-filled polymer composites. Longman, Harlow, p50

85. Hancock M, Rothon RN (1995) Principle types of particulate fillers. In: Rothon RN (ed) Particulate-filled polymer composites. Longman, Harlow, p55
86. Hancock M, Rothon RN (1995) Principle types of particulate fillers. In: Rothon RN (ed) Particulate-filled polymer composites. Longman, Harlow, p62
87. Radosta JA, Trivedi NC (1987) Talc. In: Katz HS, Milewski JV (ed) Handbook of fillers for plastics. Van Nostrand, New York, p216
88. Hancock M, Rothon RN (1995) Principle types of particulate fillers. In: Rothon RN (ed) Particulate-filled polymer composites. Longman, Harlow, p64
89. Hawley GC (1987) Flakes. In: Milewski JV, Katz HS (ed) Handbook of Reinforcements for Plastics, Van Nostrand, New York p37
90. Copeland JR (1987) Wollastonite. In: Milewski JV, Katz HS (ed) Handbook of reinforcements for plastics. Van Nostrand, New York, p125
91. Doyle M, Clemens M, Lees G, Briggs C, Day R (1994) Non-halogenated fire retardants for polypropylene. FR'94 Conference, BPF, London, p193
92. Schweitzer RA, Winterman AW (1989) Reinforcing fibres and fillers. In: Lutz JT Jr (ed) Thermoplastic polymer additives. Marcel Dekker, New York, p417
93. Dockum F Jr (1987) Fibreglass. In: Milewski JV, Katz HS (ed) Handbook of reinforcements for plastics. Van Nostrand, New York, p233
94. Jones FR (1994) Glass fibres. In: Jones FR (ed) Handbook of polymer-fibre composites. Longman, Harlow, p38
95. Katz HS (1987) Solid spherical fillers. In: Katz HS, Milewski JV (ed) Handbook of fillers for plastics. Van Nostrand, New York, p429
96. Watson SK (1987) Synthetic silicas. In: Katz HS, Milewski JV (ed) Handbook of fillers for plastics. Van Nostrand, New York, chap 9, p165
97. Sobolev I, Woycheshin EA (1989) Alumina trihydrate. In: Katz HS, Milewski JV (ed) Handbook of fillers for plastics. Van Nostrand, New York, p294
98. Hancock M, Rothon RN (1995) Principle types of particulate fillers. In: Rothon RN (ed) Particulate-filled polymer composites. Longman, Harlow, p74
99. Godfrey E, Evans KA (1987) New developments in alumina trihydrate flame retardants. Flame Retardants '87 Conference, BPF/PRI, London
100. Hancock M, Rothon RN (1995) Principle types of particulate fillers. In: Rothon RN (ed) Particulate-filled polymer composites. Longman, Harlow, p77
101. Miyata S (1976) US Patent No 4,098,762
102. Eisler J (1990) Czechoslovak Patent No 275,256
103. Elsner D, Rothon RN (1995) Patent Application No WO 95/19935
104. Skubla P (1995) European Patent Application No 0 631,984 A2
105. Meier A (1990) Patent Application No WO 90/13516
106. Horose T, Miyata S (1990) European Patent Application No 0 356 139 A1
107. Rothon RN (1994) Acicular particulate reinforcements. In: Jones FR (ed) Handbook of polymer-fibre composites. Longman, Harlow, p4
108. Birchall JD, Hutchinson J (1978) British Patent No. 1,603,300
109. Touval I (1987) Antimony oxide. In: Katz HS, Milewski JV (ed) Handbook of fillers for plastics. Van Nostrand, New York, p279

Received: March 1998

Adhesion and Surface Modification

Béla Pukánszky[1], Erika Fekete[2]

[1] Technical University of Budapest, Department of Plastics and Rubber Technology, H-1521 Budapest, P.O. Box 92, Hungary
E-mail: pukanszky.mua@chem.bme.hu
[2] Central Research Institute for Chemistry, Hungarian Academy of Sciences, H-1525 Budapest, P.O. Box 17, Hungary

This review emphasizes the role of interactions in particulate filled composites. In an introductory section a general view is given about the factors influencing composite properties. Two basic type of interactions must be considered: particle/particle and matrix/filler interaction. The effect of the former is detrimental to composite properties, it decreases strength and impact resistance. The occurrence and extent of aggregation is determined by the relative adhesion and shear forces during homogenization. Due to matrix/filler interaction an interphase forms spontaneously in the composite with properties different from those of both components. The amount and characteristics of the interphase strongly influence composite properties. The strength of adhesion between the components can be characterized by thermodynamic quantities, mainly by the reversible work of adhesion. The most important techniques used for the estimation of the strength of interaction and the properties of the interphase are briefly reviewed. The modification of interactions is achieved through the surface treatment of the filler. Surface treatments are divided into four arbitrary groups and are discussed accordingly, i.e. non-reactive and reactive treatment, application of functionalized polymers and introduction of a soft interlayer around the particles. The practical relevance of interactions and their modification is also mentioned in the last section.

Keywords: aggregation, interfacial interaction, reversible work of adhesion, wettability, matrix-filler interaction, surface treatment, interphase, surfactant, coupling agent, elastomer interlayer

List of Abbreviations and Symbols

a	surface area of an adsorbed molecule
d	particle diameter
f	conversion factor
$\dot{h}$	separation rate of particles
h_0	initial distance of particles
j_0	correction factor for pressure difference in IGC
k	constant
l	thickness of the interphase
n	number of molecules taking part in the interaction
p	vapor pressure
r	particle radius
r_a	equivalent radius, $r_a=r_1r_2/(r_1+r_2)$
t_0	reference time
t_r	retention time
A_f	specific surface area of the filler
AN, DN	Guttman's acceptor and donor numbers
B_y	constant related to stress transfer

C	geometric constant
C_1, C_2	geometric constants in debonding
C_a, C_b	Drago's acid base constants (covalent)
E_a, E_b	Drago's acid base constants (electrostatic)
F	flow rate of carrier gas in IGC
F_a	adhesive force
F_1, F_2	components of capillary force
F_h	hydrodynamic force
ΔH^{ab}	free enthalpy change due to interaction
N	Avogadro´s number
R	universal gas constant
S_{AB}	wettability
T	absolute temperature
V_N	retention volume
W_{AB}	reversible work of adhesion
W_{AB}^d, W_{BA}^p	dispersion and polar components of W_{AB}
W_{AB}^{ab}	acid base component of W_{AB}
γ_A, γ_B	surface tension of the components
γ_{AB}	interfacial tension
γ_A^d, γ_A^p	dispersion and polar components of surface tension for component A
γ_B^d, γ_B^p	dispersion and polar components of surface tension for component B
γ_{LV}	surface tension of liquids
γ_{SV}	surface tension of solids in a vapor
γ_S	surface tension of solids
γ_{SL}	solid–liquid interfacial tension
$\dot{\gamma}$	shear rate
η	viscosity of the media
θ	contact angle
ξ	geometric constant
π_e	spreading pressure
ρ_f	density of filler
ρ_1, ρ_2	geometric constants
σ_{yi}	yield stress of the interface
σ^D	debonding stress
σ^T	thermal stress
σ_{y0}, σ_y	yield stress of the matrix and the composite
φ_f	volume fraction of filler
φ_f^{max}	maximum packing fraction
AES	Auger electron spectroscopy
AMPTES	(3-aminopropyl)triethoxysilane
DRIFT	Diffuse reflectance infrared spectroscopy

FTIR	Fourier transform infrared spectroscopy
HDPE	high density polyethylene
IGC	inverse gas chromatography
MA-PP	maleinated polypropylene
PE	polyethylene
PP	polypropylene
PVC	poly(vinyl chloride)
pPVC	plasticized PVC
SIMS	Secondary ion mass spectrometry
SPTES	(3-stearyloxypropyl)triethoxysilane
XPS, ESCA	X-ray photoelectron spectroscopy

1 Introduction

The growth rate of the use of particulate filled polymers is very fast in all fields of application [1]. Household articles and automotive parts are equally prepared from them. In the early stages, the sole reason for the introduction of fillers was to decrease the price of the polymer. However, as a result of filling all properties of the polymer change, a new polymer is in fact created. Some characteristics improve, while others deteriorate, and properties must be optimized to utilize all potentials of particulate filling. Optimization must include all aspects of the composites from component properties, through structure and especially interactions.

The characteristics of particulate filled polymers are determined by the properties of their components, composition, structure and interactions [2]. These four factors are equally important and their effects are interconnected. The specific surface area of the filler, for example, determines the size of the contact surface between the filler and the polymer, thus the amount of the interphase formed. Surface energetics influence structure, and also the effect of composition on properties, as well as the mode of deformation. A relevant discussion of adhesion and interaction in particulate filled polymers cannot be carried out without defining the role of all factors which influence the properties of the composite and the interrelation among them.

This chapter focuses its attention on the discussion of the most relevant questions of interfacial adhesion and its modification in particulate filled polymers. However, because of the reasons mentioned in the previous paragraph, the four factors determining the properties of particulate filled polymers will be discussed in the first section. Interactions can be divided into two groups, particle/particle and matrix/filler interactions. The first is often neglected although it may determine the properties of the composite and often the only reason for surface modification is to hinder its occurrence. Similarly important, but a very contradictory question is the formation and properties of the interphase; a separate section will address this question. The importance of interfacial adhesion is also shown by the numerous efforts to modify them. The most important surface modification techniques are discussed in a separate section, as will the effect which can be achieved by using them, their mechanism, optimization and eventual use. Finally, the practical relevance of interactions and their modification will also be mentioned at the end of the chapter.

2
Factors Determining the Properties of Particulate Filled Polymers

As was mentioned above, four main factors determine the properties of particulate filled polymers: characteristics of the components, composition, structure and component interactions. All four are equally important and must be adjusted to achieve optimum properties and economics.

2.1
Component Properties

The properties of the matrix also strongly influence the effect of the filler on composite properties. The reinforcing effect of the filler increases with decreasing matrix stiffness. In elastomers true reinforcement takes place, both stiffness and strength increase [3]. This is demonstrated well by Fig. 1, where tensile yield stress of $CaCO_3$ composites are plotted against composition for two different matrices. LDPE is reinforced by the filler, while yield stress of PVC continuously decreases with increasing filler content [4]. For the sake of easier comparison the data were plotted on a relative scale, related to the yield stress of the matrix. The direction of the change in yield stress or strength is determined by the relative load-bearing capacity of the components [5,6]. In weak matrices the filler carries a significant part of the load, it reinforces the polymer. The extent of stress transfer depends on the strength of the adhesion between the components. If this is weak the separation of the interfaces takes place even under the effect of small external load [7,8]. With increasing matrix stiffness the effect of interaction becomes dominant, the relative load-bearing capacity of the components is determined by this factor. In a stiffer matrix larger stresses develop around the inclusions and the probability of dewetting increases. In such matrices dewetting is usually the dominating micromechanical deformation process.

The structure of crystalline polymers may be significantly modified by the introduction of fillers. All aspects of the structure change on filling, crystallite and spherulite size, as well as crystallinity, are altered as an effect of nucleation [9]. A typical example is the extremely strong nucleation effect of talc in polypropylene [10,11], which is demonstrated also in Fig. 2. Nucleating effect is characterized by the peak temperature of crystallization, which increases significantly on the addition of the filler. Elastomer modified PP blends are shown as a comparison; crystallization temperature decreases in this case. Talc also nucleates polyamides. Increasing crystallization temperature leads to an increase in lamella thickness and crystallinity, while the size of the spherulites decreases on

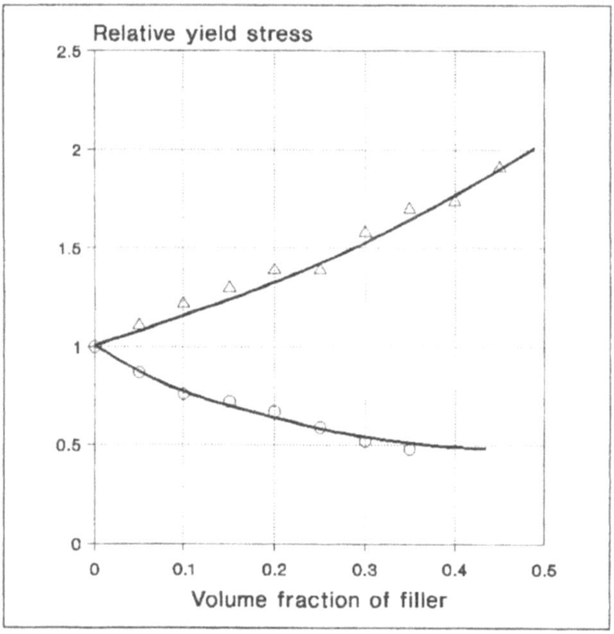

Fig. 1. Effect of matrix properties on the tensile yield stress of particulate filled composites. Symbols: (○) PVC, (△) LDPE; filler: $CaCO_3$, r=1.8 μm

nucleation. Increasing nucleation efficiency results in increased stiffness and decreased impact resistance [12,13].

Numerous filler characteristics influence the properties of composites [14,15]. Chemical composition and, in particular, purity of the filler both have a direct and an indirect effect on its application possibilities and performance. Traces of heavy-metal contamination decrease stability. Insufficient purity leads to discoloration, high purity $CaCO_3$ has the advantage of a white color, while the grey shade of talc filled composites excludes them from some fields of application.

The mechanical properties of PP composites containing non-treated fillers are determined mainly by their particle characteristics, i.e. particle size and particle size distribution. Strength, sometimes modulus increase, deformability and impact strength decrease with decreasing particle size. Particle size in itself, however, is not sufficient for the characterization of any filler, the knowledge of the particle size distribution is equally important [16]. In the case of large particles the volume in which stress concentration is effective is said to increase with particle size and the strength of matrix/filler adhesion also depends on it. The other end of the particle size distribution, i.e. the amount of small particles, is equally important. Aggregation tendency of fillers increases with decreasing particle size. Extensive aggregation, however, leads to insufficient homogeneity,

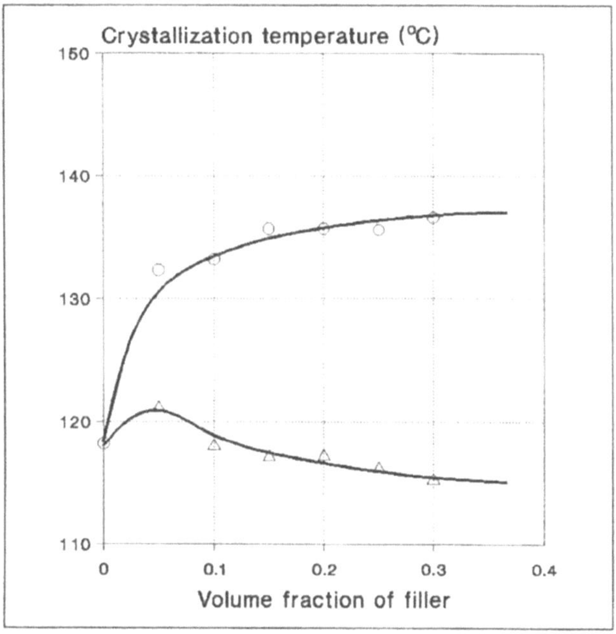

Fig. 2. Nucleation effect of talc in PP. Symbols: (○) talc, (△) EPR

rigidity and lower impact strength. Aggregated filler particles act as crack initiation sites under dynamic loading conditions.

The specific surface area of fillers is closely related to their particle size distribution; however, it also has a direct impact on composite properties. Adsorption of both small molecular weight additives, and also that of the polymer is proportional to the size of the matrix/filler interface [14]. Adsorption of additives may change stability, while matrix/filler interaction significantly influences mechanical properties, first of all yield stress, tensile strength and impact resistance [5,6].

Anisotropic particles reinforce polymers, the effect increasing with the anisotropy of the particle. In fact, fillers and reinforcements are very often differentiated by their degree of anisotropy (aspect ratio). Plate-like fillers like talc and mica reinforce polymers more than spherical fillers and the influence of glass fibers is even stronger [17]. Anisotropic fillers orientate during processing further enhancing their reinforcing effect, which depends very much also on the distribution of orientation.

The hardness of the filler has a strong effect on the wear of the processing equipment, but this is also influenced by the size and shape of the particles, the composition, viscosity, and speed of processing etc. [16].

Surface free energy (surface tension) of the fillers determines both matrix/filler and particle/particle interaction. The former has a pronounced effect on the mechanical properties, the latter determines aggregation. Both interactions can be modified by surface treatment.

The thermal properties of fillers differ significantly from those of thermoplastics. This has a beneficial effect on productivity and processing. Decreased heat capacity and increased heat conductivity reduce cooling time [16]. Changing thermal properties of the composites result in a modification of the skin-core morphology of crystalline polymers and thus in the properties of injection molded parts as well. Large differences in the thermal properties of the components, on the other hand, lead to the development of thermal stresses, which also influence the performance of the composite under external load.

Although a number of filler characteristics influence composite properties, particle size, specific surface area, and surface energetics must again be mentioned here. All three also influence interfacial interactions. In the case of large particles and weak adhesion, the separation of the matrix/ filler interface is easy, debonding takes place under the effect of a small external load. Small particles form aggregates which cause a deterioration in the mechanical properties of the composites. Specific surface area, which depends on the particle size distribution of the filler, determines the size of the contact surface between the polymer and the filler. The size of this surface plays a crucial role in interfacial interactions and the formation of the interphase.

2.2 Composition

Composition, i.e. filler content of composites, may change over a wide range. The effect of changing composition on composite properties is clearly seen in Fig. 1. The interaction of various factors determining composite properties is also demonstrated by this figure, the same property may change in a different direction as a function of matrix characteristics, but interfacial adhesion may also have a similar effect. The goal of filling is either to decrease price or to improve properties, e.g. stiffness, dimensional stability etc. These goals require the introduction of the largest possible filler content, which, however, may lead to the deterioration of other properties. An optimization of properties must be carried out during the development of composites. Numerous models are available which describe the composition dependence of the various properties of the composites [2,15,17]. With the help of these models composite properties may be predicted and an optimization can be carried out. In spite of their importance these questions will not be discussed in this chapter.

2.3 Structure

The structure of particulate filled polymers seems to be simple, homogeneous distribution of particles is assumed in most cases. This, however, rarely occurs and often special, particle related structures develop in the composites. The most important of these are aggregation and orientation of anisotropic filler particles. The first is related to the interactions acting in a particulate filled polymer, thus it will be discussed in detail in a subsequent section (see Sect. 3.1).

2.4 Interfacial Interactions

Particle/particle interactions induce aggregation, while matrix/filler interaction leads to the development of an interphase with properties different from those of both components. Both influence composite properties significantly. Secondary, van der Waals forces play a crucial role in the development of these interactions. Their modification is achieved by surface treatment. Occasionally reactive treatment is also used, although its importance is smaller in thermoplastics than in thermoset matrices. In the following sections of this chapter attention is focused on interfacial interactions, their modification and on their effect on composite properties.

3 Interactions

As was mentioned in the previous section two types of interactions must be considered in particulate filled polymers: particle/particle and matrix/filler interaction. The first is often neglected even by compounders, in spite of the fact that its presence may cause composite properties to deteriorate significantly especially under the effect of dynamic loading conditions [18]. Many attempts have been made to change both interactions by the surface treatment of the filler, but the desired effect is often not achieved due to improper use of incorrect ideas.

3.1 Particle/Particle Interaction – Aggregation

Aggregation is a well-known phenomenon in particulate filled composites. Experience has shown that the probability of aggregation increases with decreasing particle size of the filler. The occurrence and extent of aggregation are determined by the relative magnitude of the forces which bind together the particles, on the one hand, or try to separate them on the other. Particulate filled polymers are prepared by melt mixing of the components, thus the major attractive and separating forces must be considered under these conditions. Ess and Hornsby [19] carried out a detailed investigation of particle characteristics and processing conditions on the dispersive mixing in particulate filled PA and PP. They listed the principal adhesive forces, which in increasing order of adhesive strength are mechanical interlocking, electrostatic forces, van der Waals forces, liquid bridges, and solid bridging. Particles are separated mainly by shear and elongational forces during homogenization.

Since aggregation is also an important phenomenon in other areas (pigments, paints, powder handling, etc.) numerous studies deal with the interaction of particles [20]. When two bodies enter into contact they are attracted to each other. The strength of adhesion between the particles is determined by their size and surface energy [21,22], i.e.:

$$F_a = \frac{3}{2}\pi W_{AB} r_a \tag{1}$$

where F_a is the adhesive force between the particles, W_{AB} is the reversible work of adhesion and $r_a=r_1r_2/(r_1+r_2)$, an effective radius, when the size of the two interacting particles is different.

In the presence of fluids, i.e. in suspensions, and also in the polymer melt during homogenization, further forces act between the particles. Adam and Edmondson [22] specify two attractive forces, depending on the extent of wetting of the particles. In the case of complete wetting, viscous force (F_v) acts between the particles, which are separated from each other at a constant rate. F_v can be expressed as:

$$F_v = 3\pi\eta r^2 \frac{\dot{h}}{2h_0} \tag{2}$$

where η is the viscosity of the fluid, $\dot{h}$ the separation rate and h_0 is the initial distance of the particles. The rate of separation can be approximated by:

$$\dot{h} = d\dot{\gamma} \tag{3}$$

where d is the diameter of the particles. The viscous force might have some importance during the homogenization of the composites.

If the particles are wetted only partially by the fluid (melt), liquid bridges form and capillary forces develop between them. These can be divided into two parts: that related to surface tension:

$$F_1 = 2\pi C\gamma_{LV}\cos\zeta \tag{4}$$

and that to the hydrostatic component:

$$F_2 = \pi C^2 \gamma_{LV}\left(\frac{1}{\rho_1} - \frac{1}{\rho_2}\right) \tag{5}$$

where γ_{LV} is the surface tension of the liquid and C, ζ, ρ_1, ρ_2 are geometric parameters.

Balachandran [23] analyzed the role of electrostatic forces in the adhesion between solid particles and surfaces. According to the available information this role is not completely clear, the results are contradictory. Electrostatic forces can be significant in the case of polymer and semi-conductor particles. These forces have four main types: Coulomb, image charge, space charge and dipole forces.

The magnitude of all four forces is around 10^{-7}–10^{-8} N, which is significantly smaller than that of the other forces acting between the particles. However, if the particles are small, these forces can be larger than their weight. These forces may play an important role during the approach of particles, but they are dissipated on contact and van der Waals or other adhesive forces become dominant. The effect of electrostatic forces is important only in the case of charged particles or surfaces.

The number of forces separating the particles is smaller. Repulsive forces act between particles with the same electrostatic charge. The mixing of fluids leads to the development of shear forces, which try to separate the particles. The maximum hydrodynamic force acting on spheres in a uniform shear field can be expressed as [22,24]:

$$F_h = -6.12 \pi \eta r^2 \dot{\gamma} \quad (6)$$

The adhesive and separating forces acting between the particles are summarized in Table 1. Only their magnitude is indicated, since it is difficult to determine them exactly. Most of the values are given for 20 μm large particles, adhesion and hydrodynamic forces were calculated for PP/$CaCO_3$ composites. In the case of viscous forces the initial distance of the particles, while for capillary forces the filling volume and separation distance, determine the exact value of the force. The table clearly shows that the balance of adhesive and hydrodynamic forces determines the extent of aggregation during the homogenization of a polymer composite. This is somewhat in contradiction with the statement of Ess and Hornsby [19] who found capillary forces more important than van der Waals interactions.

Table 1. Forces Acting During the Homogenization of Particulate Filled Composites Calculated for Particles of 20 μm Diameter

Direction	Character	Type	Magnitude (N)	Reference
		Force		
Adhesive	**Adhesive**		**10^{-4}**	Eq. 1
	Viscous		10^{-5}–10^{-6}	Eqs. 2–3
	Capillary	Surface tension	10^{-5}–10^{-6}	22[a]
		Hydrostatic	10^{-5}–10^{-6}	22[a]
	Electrostatic	Coulomb	10^{-7}	23
		Dipole	10^{-8}	23
		Image charge	10^{-7}	23
		Space charge	10^{-8}	23
Separating	**Hydrodynamic**		**10^{-3}**	Eq. 6
	Electrostatic		10^{-8}	23

[a] For particles of 100 μm diameter

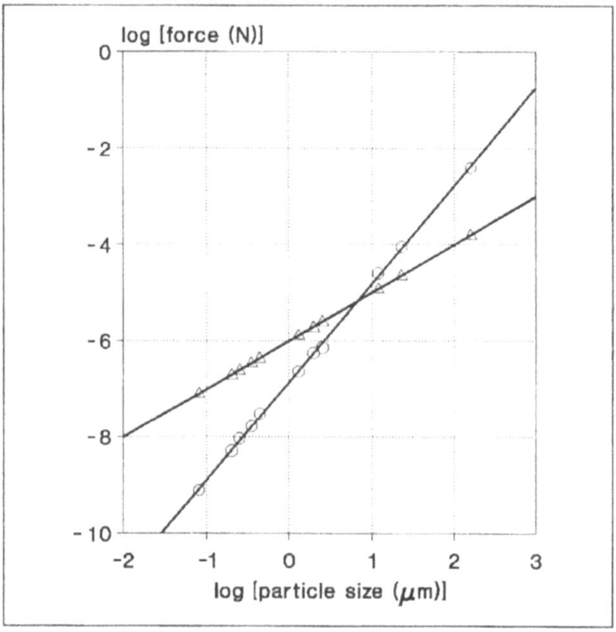

Fig. 3. Effect of filler particle size on the adhesion and shear forces acting during the homogenization of PP/$CaCO_3$ composites. Symbols: (○) shear, (△) adhesion [25]

Both adhesive and hydrodynamic forces depend on the size of the particles. The two forces were calculated for $CaCO_3$ fillers of various particle sizes homogenized in a PP matrix. The results are presented in Fig. 3. At a certain particle size adhesion exceeds shear forces, aggregation of the particles takes place in the melt. Since commercial fillers have a relatively broad particle size distribution, most fillers show some degree of aggregation and the exact determination of the particle size, or other filler characteristics where aggregation appears, is difficult. Experiments carried out with 11 different $CaCO_3$ showed this limit to be around 6 m^2/g specific surface area [25].

Since the relative magnitude of adhesive and shear forces determine the occurrence and extent of aggregation in a composite the ratio of the two forces may give information about the possibilities to avoid or decrease it, i.e.:

$$\frac{F_a}{F_h} = k \frac{W_{AB}}{\eta \dot{\gamma} r} \tag{7}$$

An increase in shear rate and particle size will lead to decreased aggregation. Naturally both can be changed only in a limited range since excessive shear leads to degradation, while large particles easily separate from the matrix under the effect of external load as was pointed out in Sect. 2.1. Equation 7 also explains the fact that a filler with a smaller particle size can be incorporated into unplasticized PVC, the high viscosity of this polymer results in better dispersion. According to Eq. 7, lower reversible work of adhesion also results in a decreased aggregation tendency. Non-reactive surface treatment invariably leads to a decrease in surface tension and W_{AB}, thus to decreased aggregation, improved processability and mechanical properties.

The presence of aggregates is almost invariably detrimental to the properties of composites. Figure 4 demonstrates this effect. The strength of PP/$CaCO_3$ composites initially increases with increasing specific surface area of the filler, but it strongly decreases when aggregation takes place at small particle sizes. The effect is even more pronounced in the case of impact properties, the fracture resistance of composites containing aggregated particles drastically decreases with increasing number of aggregates [9,18].

3.2
Matrix/Filler Interaction – Interfacial Adhesion

The polymers used as matrices in thermoplastic composites as well as fillers have the most diverse physical and chemical structures, thus a wide variety of interactions may form between the two components. Two boundary cases of inter-

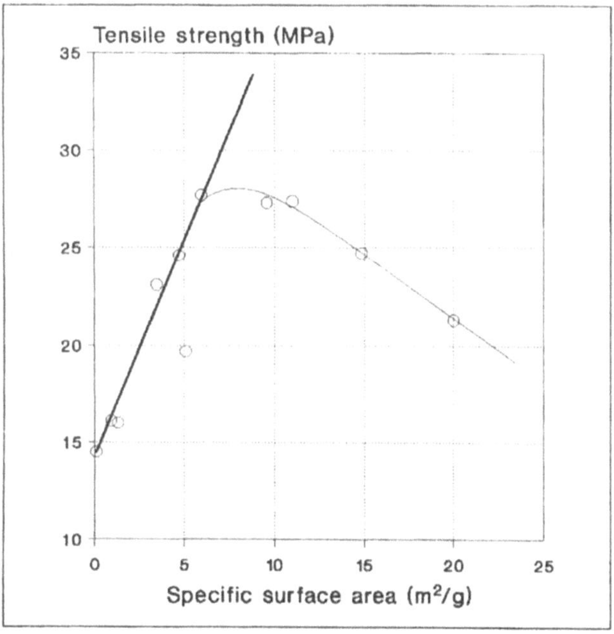

Fig. 4. Effect of aggregation on the strength of PP/$CaCO_3$ composites [25]

actions can be distinguished: covalent bonds, which rarely form spontaneously, but can be created by special surface treatments and zero interaction, which does not exist in reality, since at least secondary, van der Waals forces always act between the components.

In practice the strength of the interaction is somewhere between the two boundary cases. Interaction between two surfaces in contact with each other can be created by primary or secondary bonds. The most important primary forces are the ionic, covalent and metallic bonds. The bonds formed by these forces are very strong, their strength is between 60–80 kJ/mol for covalent and 600–1200 kJ/mol for ionic bonds [26]. The secondary bonds are created by van der Waals forces, i.e. by dipole–dipole (Keesom), induced dipole (Deby) and dispersion (London) interaction. The strength of these interactions is much lower, it is between 20 and 40 kJ/mol [26]. H-bond forms a transition between the two groups of interactions, both in character and strength. Besides the attractive forces created by the above mentioned secondary forces, repulsive forces also act between the interacting surfaces due to the interaction of their electron fields. The final distance of the atoms is determined by the equilibrium of the attractive and repulsive forces. As mentioned before, zero interaction does not exist, one of the forces always act between the surfaces in contact, but besides these forces numerous other factors may influence interaction, i.e. the strength of the adhesive bond.

The number of these factors and their complicated relationship is clearly shown by the numerous theories which attempt to describe the phenomenon of adhesion. These theories represent different approaches to the problem, describe one interaction well, but usually do not offer a general solution [27,28].

The theory of *mechanical interlocking* explains adhesive bonding with the physical coupling of surface irregularities, roughness. It can be applied for the solution of problems emerging in the textile and paper industry, but cannot describe adhesive interaction of smooth surfaces like in the case of glass.

According to the *theory of interdiffusion* adhesion is caused by the mutual diffusion of the molecules of the interacting surfaces. This theory can be applied for the description of interaction in polymer blends, but its use is limited when solid surfaces are in contact (aggregation, filler/matrix interaction).

Derjaugin [29] explains the significant interaction which sometimes is created between polymers and metals by *electrostatic interaction*. According to his reasoning the polymer and the thin metallic film layer correspond to an electric double layer, which forms by charge transfer between the surfaces. The significance of this theory in particulate filled polymers is very limited.

Adhesion is created by primary and secondary forces according to the theory of *adsorption interaction*. This theory is applied the most widely for the description of interaction in particulate filled or reinforced polymers [30]. The approach is based on the theory of contact wetting and focuses its attention mainly on the influence of secondary forces. Accordingly, the strength of the adhesive bond is assumed to be proportional to the reversible work of adhesion (W_{AB}), which is necessary to separate two phases with the creation of two new surfaces.

The Dupré equation relates W_{AB} to the surface (γ_A and γ_B) and interfacial (γ_{AB}) tension of the components, i.e.:

$$W_{AB} = \gamma_A + \gamma_B - \gamma_{AB} \tag{8}$$

One of the major problems of the application of the theory to solids is the determination of the interfacial tension. For such systems it cannot be determined by direct measurements, it is usually derived from thermodynamic calculations. Good and Girifalco [31] developed the first theory for the calculation of γ_{AB}, but because it contained an adjustable parameter it did not gain practical use. The most widely accepted solution was suggested by Fowkes [32,33]. He assumed that surface tension can be divided as shown in Eq. 9:

$$\gamma = \gamma^d + \gamma^p + \gamma^i + \gamma^H + \gamma^\pi + \gamma^m + \dots \tag{9}$$

where the components noted by the superscripts d, p, i, H, π and m correspond to interactions created by dispersion, dipole–dipole, induced-dipole, hydrogen, π and metallic bonds. Only dispersion forces act between apolar surfaces and in such cases interfacial tension can be expressed by Eq. 10:

$$\gamma_{AB} = \gamma_A + \gamma_B - 2(\gamma_a^d \ \gamma_b^d)^{1/2} \tag{10}$$

Although it was assumed that Eq. 10 is also valid when an apolar material enters into interaction with a polar one, in practice polar surfaces interact with each other more often. Several attempts were made to generalize the correlation of Fowkes for such cases and the geometric mean approximation gained the widest acceptance. This considers only the dispersion and a polar component of the surface tension, but the latter includes all polar interactions [34]. Thus interfacial interaction can be calculated as follows:

$$\gamma_{AB} = \gamma_A + \gamma_B - 2(\gamma_A^d \gamma_B^d)^{1/2} - 2(\gamma_A^p \gamma_B^p)^{1/2} \tag{11}$$

The surface tension of two thermoplastics and three fillers are listed in Table 2. Large differences can be observed both in the dispersion, but especially in the polar component. The surface tension of the majority of polymers is in the same range, in fact between that of PP and PMMA. Those listed in Table 2 represent the most important particulate fillers, and also reinforcements used in practice, since clean glass fibers possess similar surface tensions to SiO_2. Surface treatment lowers the surface tension of fillers significantly (see Sect. 6.1).

Although Eq. 11 tries to take into account the effect of polar interactions, the role of acid/base interactions in adhesion became clear and theories describing them have been more and more accepted [35]. The boundary case of such interactions is the formation of covalent bonds between the surfaces. Such interactions cannot be described by Eq. 11. As a consequence Fowkes [36] suggested that the reversible work of adhesion should be defined as:

$$W_{AB} = W_{AB}^d + W_{AB}^{ab} + W_{AB}^p \tag{12}$$

Table 2. Surface Tension of Selected Polymers and Fillers; Dispersion (γ^d) and Polar (γ^p) Components

Material	Surface tension (mJ/m^2) γ^d	γ^p	γ
PP[a]	32.5	0.9	33.4
PMMA[a]	34.3	5.8	40.1
$CaCO_3$[b]	54.5	153.4	207.9
talc[c]	49.3	90.1	139.4
SiO_2[c]	94.7	163.0	257.7

[a] Contact angle, [b] IGC, [c] Gravimetric measurement

where W_{AB}^{ab} is the part of the reversible work of adhesion created by acid/base interactions. According to Fowkes the polar component can be neglected, i.e. $W_{AB}^{p} \sim 0$, thus W_{AB} can be expressed as [36,37]:

$$W_{AB} = 2\sqrt{\gamma_A^d \gamma_B^d} + n f \Delta H^{ab} \tag{13}$$

where ΔH^{ab} is the change in free enthalpy due to acid/base interactions, n is the number of moles taking part in the interaction on a unit surface and f is a conversion factor which takes into account the difference between free energy and free enthalpy (f~1).

The value of ΔH^{ab} can be calculated from the acid-base constants of the interacting phases by using the theory of Drago [38] or Guttman [39]. Drago [38] suggested two constants for each material and divided the compounds into acids and bases. ΔH^{ab} can be calculated as:

$$-\Delta H^{ab} = C_A C_B + E_A E_B \tag{14}$$

where subscripts A and B represent acid and base, C expresses the covalent, while E the electrostatic character of the interaction.

Guttman [39] characterized materials by a donor (DN) and acceptor (AN) number, which indicate the acidity or basicity of the material. According to this theory:

$$-\Delta H^{ab} = \frac{AN \cdot DN}{100} \tag{15}$$

In spite of the fact that theoretically the approach of Drago is more sound, the use of the AN and DN numbers of Guttman is more accepted. These numbers can also characterize amphoteric materials and the available data base is much larger for this theory. Schreiber [30], for example, lists these parameters for a large number of polymers and other components used in polymer systems.

The strength of the adhesive bond is described acceptably by the reversible work of adhesion values calculated by the above theory in most cases. Often, especially in apolar systems, a close correlation exists between W_{AB} and the macroscopic properties of the composite even in particulate filled polymers. The dependence of composite strength on W_{AB} in PP/$CaCO_3$ composites, which is presented in Fig. 5, strongly supports this statement. However, the bond strengths determined with various mechanical tests are usually much lower than the values derived from theoretical calculations. This difference can be traced back to the imperfections of the interface, which lead to the decrease in bond strength. The different elastic properties of the components usually result in the development of stress concentrations, which can cause either the fracture of the interface or lead to the cohesive failure of one of the phases. This observation led to the development of the theory of *weak boundary layers* [40].

In spite of the imperfections of the approach, the reversible work of adhesion can be used for the characterization of matrix/filler interactions in particulate filled polymers. Debonding is one of the dominating micromechanical processes in these materials. Stress analysis has shown that debonding stress (σ^D) depends on the reversible work of adhesion [8], i.e.:

$$\sigma^D = -C_1 \sigma^T + \sqrt{\frac{C_2 W_{AB}}{R}} \tag{16}$$

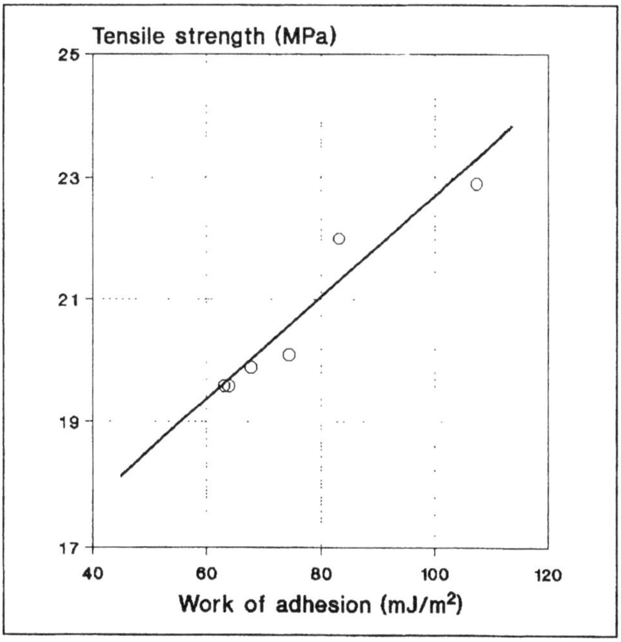

Fig. 5. Effect of interaction on the strength of PP/$CaCO_3$ composites; filler: $CaCO_3$, φ_f=0.2, r=0.9 μm

where σ^T is thermal stress, R the radius of the particle and C_1 and C_2 are factors determined by the geometric conditions of debonding. The mechanism of deformation depends on the strength of the interaction, in the case of weak adhesion debonding takes place and yield stress decreases with increasing filler content, while strong adhesion leads to the yielding of the matrix polymer or to other micromechanical deformation processes [4,41]. Figure 6 sufficiently supports both the influence of adhesion (W_{AB}) on composite properties and the dominating role of debonding in particulate filled polymers.

Since the effective range of the forces creating the adhesive bond is much smaller than the irregularities of the surface, appropriate wetting is an important condition of interaction. Wetting has thermodynamic and kinetic conditions. The thermodynamic conditions are usually satisfied since polymers having relatively low surface tension completely wet the high energy surfaces of fillers (see Table 2) [42]. However, the kinetic conditions are more difficult to fulfill, the high viscosity melt cannot always penetrate into small crevices, narrow channels or the space inside an aggregate, thus surface defects form which finally lead to the premature failure of the composite.

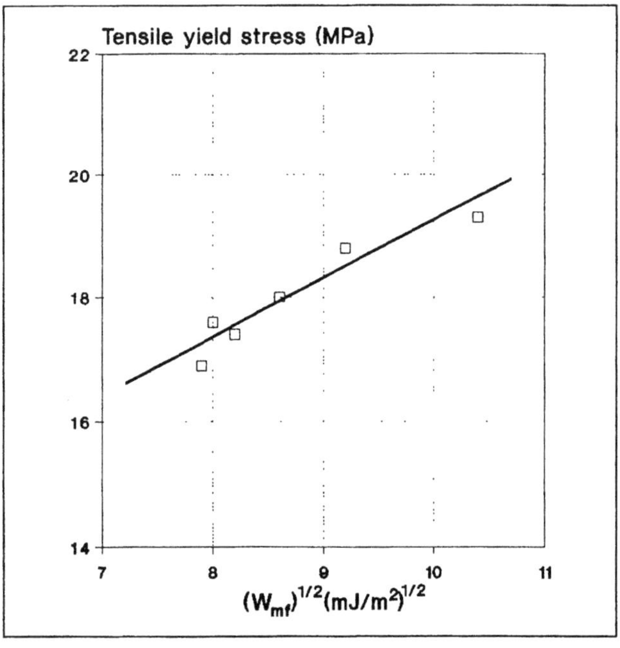

Fig. 6. Debonding of interfaces, effect of adhesion on PP/$CaCO_3$ composites; filler: $CaCO_3$, φ_f=0.2, r=1.8 μm

4
Interphase – Thickness, Properties, Role

As Table 2 shows, non-treated fillers and reinforcements have high energy surfaces. During the almost exclusively used melt mixing procedure, the forces discussed in the previous section lead to the adsorption of polymer chains onto the active sites of the filler surface. The adsorption of polymer molecules results in the development of a layer which has properties different from those of the matrix polymer [43–47]. Although the character, thickness and properties of this interlayer or interphase are much discussed topics, its existence is now an accepted fact.

4.1
Properties

In semi-crystalline polymers the interaction of the matrix and the filler changes both the structure and the crystallinity of the interphase. The changes induced by the interaction in bulk properties are reflected by increased nucleation or by the formation of a transcrystalline layer on the surface of anisotropic particles [48]. The structure of the interphase, however, differs drastically from that of the matrix polymer [49,50]. Because of the preferred adsorption of large molecules, the dimensions of crystalline units can change, and usually decrease. Preferential adsorption of large molecules has also been proved by GPC measurements after separation of adsorbed and non-attached molecules of the matrix [49,50]. Decreased mobility of the chains affects also the kinetics of crystallization. Kinetic hindrance leads to the development of small, imperfect crystallites, forming a crystalline phase of low heat of fusion [51].

Atomistic simulation of an atactic polypropylene/graphite interface has shown that the local structure of the polymer in the vicinity of the surface is different in many ways from that of the corresponding bulk. Near the solid surface the density profile of the polymer displays a local maximum, the backbone bonds of the polymer chains develop considerable parallel orientation to the surface [52]. This parallel orientation due to adsorption can be one of the reasons for the transcrystallinity observed in the case of many anisotropic filler particles.

Decreased mobility of adsorbed chains has been observed and proved in many cases both in the melt and in the solid state [52–54] and changes in composite properties are very often explained by it [52,54]. Overall properties of the interphase, however, are not completely clear. Based on model calculations the formation of a soft interphase is claimed [51], while in most cases the increased stiffness of the composite is explained by the presence of a rigid interphase [55,56]. The contradiction obviously stems from two opposing effects. Imperfection of the crystallites and decreased crystallinity of the interphase should lead to lower modulus and strength and larger deformability. Adhesion and hindered mobility of adsorbed polymer chains, on the other hand, decrease deformability and increase the strength of the interlayer.

In many models the properties of the interphase are assumed to be constant, independent of the distance from the filler surface. However, in the case of a spontaneously formed interphase, its properties must change from the surface of the filler to the homogeneous matrix. A model based on an interlayer with continuously changing properties was developed to describe the stress–strain behavior of particulate filled polymers [4,57]. Stress analysis and model calculations have indicated that changing interlayer properties significantly alter deformation and stress fields around the particles. A comparison to experimental data showed that reasonable agreement between experiments and calculation can be obtained only if the development of a hard interlayer is assumed. Further considerations proved that beside elastic properties, the yield stress of the interphase also changes continuously from the surface of the particle to the matrix [4]. The composition dependence of composite yield stress is determined by the relation of debonding stress and the yield stress of the matrix. Strong interaction leads to matrix yielding, while weak adhesion results in debonding as the dominating micromechanical deformation process [4].

4.2 Thickness

The thickness of the interphase is a similarly intriguing and contradictory question. It depends on the type and strength of the interaction and values from 10 Åto several microns have been reported in the literature for the most diverse systems [47,49,52,58–60]. Since interphase thickness is calculated or deduced indirectly from some measured quantities, it depends also on the method of determination. Table 3 presents some data for different particulate filled systems. The data indicate that interphase thicknesses determined from some mechanical properties are usually larger than those deduced from theoretical calculations or from extraction of filled polymers [49,52,59–63]. The data supply further proof for the adsorption of polymer molecules onto the filler surface and for the decreased mobility of the chains. Thermodynamic considerations and extraction experiments yield data which are not influenced by the extent of deformation. In mechanical measurements, however, deformation of the material takes place in all cases. The specimen is deformed even during the determination of modulus. With increasing deformations the role and effect of the immobilized chain ends increase and the determined interphase thickness also increases (see Table 3) [61].

The thickness of the interphase must depend on the strength of the interaction. As was pointed out previously, interaction is created by secondary, van der Waals forces. Although the range of these forces is small, the volume affected by the decreased mobility of the chains attached to the surface is much larger when the material is deformed, shown also by the larger interphase thicknesses determined by indirect, mechanical measurements (see Table 3). This volume and the thickness of the interphase can be estimated by a semi-empirical correlation de-

Table 3. Interphase Thickness in Particulate Filled Polymers Determined by Different Techniques

Matrix polymer	Filler	Method of determination	Thickness (μm)	Reference
HDPE	SiO_2	extraction	0.0036	58
HDPE	"	"	0.0036	49
PP	"	"	0.0041	49
PP	graphite	model calc.	0.001	52
PS	mica	din. mech. meas.	0.06	60
PMMA	glass	"	1.4	60
PU	polymeric	"	0.36–1.45	62,63
PP	$CaCO_3$	Young's modulus	0.012	61
PP	"	yield stress	0.15	61
PP	"	tensile strength	0.16	61

veloped for the quantitative description of the composition dependence of tensile yield stress in heterogeneous polymer systems [5]:

$$\sigma_y = \sigma_{y0} \frac{1-\varphi_f}{1+2.5\varphi_f} \exp(B_y \varphi_f) \tag{17}$$

where σ_y and σ_{y0} are the yield stress of the composite and the matrix, respectively, φ_f is the volume fraction of the filler in the composite and B is a parameter related to stress transfer between the components. The term $(1-\varphi_f)/(1+2.5\varphi_f)$ expresses the decrease of effective load-bearing cross section on filling, while $\exp(B_y\varphi_f)$ describes interaction. The parameter B_y contains the thickness of the interphase (l) and its yield stress (σ_{yi}):

$$B_y = (1 + l\rho_f A_f) \ln \frac{\sigma_{yi}}{\sigma_{y0}} \tag{18}$$

where A_f and ρ_f are the specific surface area and the density of the filler, respectively. Parameter B_y can be determined from the composition dependence of tensile yield stress and if the experiments are carried out with at least two fillers of different particle sizes l can be calculated from the results.

Interphase thicknesses are plotted as a function of W_{AB} in Fig. 7 for $CaCO_3$ composites prepared with four different matrices: PVC, plasticized PVC (pPVC), PP and HDPE. The thickness of the interphase linearly changes with increasing adhesion. The figure proves several of the points mentioned above. The reversible work of adhesion adequately describes the strength of the interaction, or at least it is proportional to it, interaction is created mostly by secondary forces and, finally, the thickness of the interphase strongly depends on the strength of interaction.

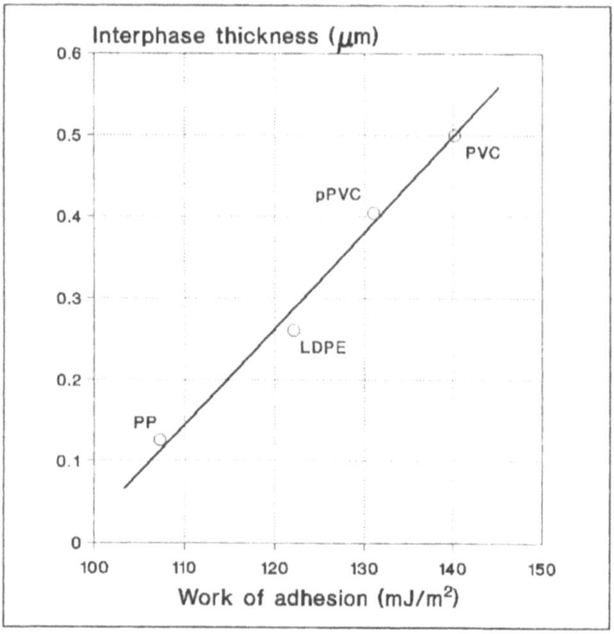

Fig. 7. Effect of interfacial interaction on the thickness of the interphase [4]

4.3 Effect on Composite Properties

The amount of polymer bonded in the interphase depends on the thickness of the interlayer and on the surface area, where the filler and the polymer are in contact with each other. The size of the interface is more or less proportional to the specific surface area of the filler, which is inversely proportional to particle size. In accordance with the above proposed explanation on the relation of the effect of immobilized polymer chains and the extent of deformation, modulus shows only a very weak dependence on the specific surface area of the filler [64].

Properties measured at significantly larger deformations, e.g. tensile yield stress or tensile strength, show much more pronounced dependence on the specific surface area of the filler than modulus [64]. Figure 8 shows that yield stresses larger than the corresponding value of the matrix can be achieved, i.e. even spherical fillers can reinforce polymers [5,6]. If adhesion is strong, yielding must be initiated at the matrix value and no reinforcing would be possible. The reinforcing effect of spherical particles can be explained only with the presence of an interphase having properties somewhere between those of the polymer and the filler [5,6] in accordance with the considerations presented above.

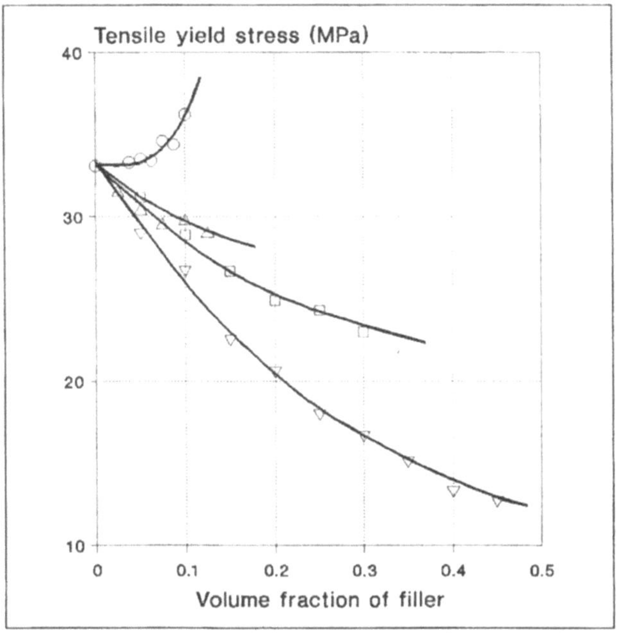

Fig. 8. Effect of the size of the interface on the yield stress of PP/$CaCO_3$ composites. Specific surface area: (▽) 0.5, (□) 3.3, (△) 16.5, (○) 200 m^2/g

5
Measurement or Estimation of Interaction

Since the thickness and properties of the interphase strongly influence the characteristics of composites and the strength of the interaction determines the dominating micromechanical deformation process, many attempts have been made to characterize them quantitatively. Many various techniques are used for this purpose, and it is impossible to give a detailed account here. As a consequence a general overview of the most often used techniques is given with a more detailed account of some specific methods which have increased importance. A more detailed description of the surface characterization techniques can be found in a recent monograph by Rothon [15].

5.1
Composition of the Interphase

The composition of spontaneously formed interphases is basically the same as that of the matrix. It is difficult, if not impossible, to detect any changes in it as an effect of interaction. Since the interlayer is very thin, microscopic techniques cannot detect it and due to the small volume involved spectroscopic techniques cannot characterize it either.

The most information about a spontaneously formed interphase is obtained by atomistic simulation [52]. Such calculations give information about the density profile in the vicinity of the surface, the orientation of the molecules and even the properties of the interphase. Unfortunately, a comparison of the results of such theoretical calculations to experimental data is difficult, or at least lacking at the moment.

Spectroscopic techniques are extremely useful for the characterization of filler surfaces treated with surfactants or coupling agents in order to modify interactions in composites. Such an analysis makes possible the study of the chemical composition of the interlayer, the determination of surface coverage and possible coupling of the filler and the polymer. This is especially important in the case of reactive coupling, since, for example, the application of organofunctional silanes may lead to a complicated polysiloxane interlayer of chemically and physically bonded molecules [65]. The description of the principles of the techniques can be found elsewhere [15,66–68], only their application possibilities are discussed here.

X-ray photoelectron spectroscopy (XPS) gives a spectrum where the peaks are highly element specific, allowing direct elemental analysis. The oxidation state of elements can be determined from the spectrum, while chemical shifts give information about functionality and substitution for organic compounds. Secondary ion mass spectrometry (SIMS) yields charged fragments of molecules covering the surface. Dynamic SIMS is useful for trace element analysis, while static SIMS is used for the structural characterization of surfaces. Although a handbook of standard spectra of typical molecules used for surface treatment is available [69], the evaluation of the spectra may be difficult. Auger electron spectroscopy (AES) is another potential technique for surface analysis, but according to Ashton and Briggs [70] technical problems arise during measurement due to beam damage to organic materials and charging of highly insulating powders.

Diffuse reflectance infrared spectroscopy (DRIFT) gives valuable information about the chemical and physical changes taking place on a filler surface on adsorption. The analysis can be carried out on powders, which is a great advantage of the technique. Changes in the characteristic vibrations indicate chemical reactions, while a shift in an absorption band shows physico-chemical interaction. The spectrum is recorded in Kubelka–Munk units which is proportional to concentration. The FTIR spectra of a (3-stearyloxipropyl) triethoxysilane coupling agent (SPTES) and the DRIFT spectra of $CaCO_3$ filler covered with the compound is compared in Fig. 9. Polycondensation of the silane takes place on the surface of the filler, the different structure of the adsorbed layer and the original coupling agent is obvious from the comparison. Further analysis even gives information about the structure of the polysiloxane layer.

Adsorption–desorption techniques are often used for the study of surface treatment or modeling of polymer/filler interaction. Gravimetric techniques or,

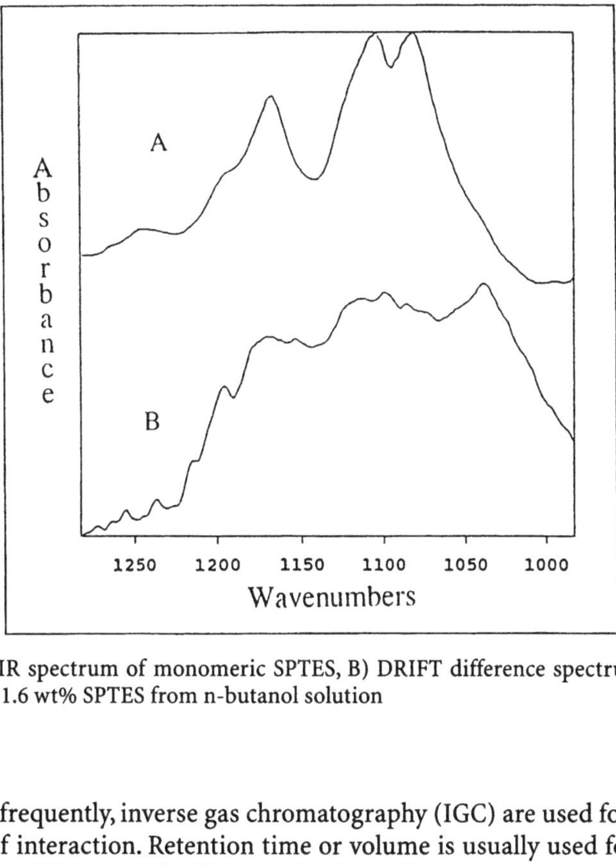

Fig. 9. A) FTIR spectrum of monomeric SPTES, B) DRIFT difference spectrum of $CaCO_3$ treated with 1.6 wt% SPTES from n-butanol solution

even more frequently, inverse gas chromatography (IGC) are used for the determination of interaction. Retention time or volume is usually used for the characterization of the strength of interaction [71,72].

5.2
Measurement of Thermodynamic Parameters

The direct determination of matrix/filler interaction is difficult, indirect techniques are used in most cases. These employ the principles discussed in Sect. 3.2. The surface tension of the components and interfacial tension or acid/base interaction parameters must be known in order to determine the reversible work of adhesion. Adsorption–desorption techniques, which use small molecular weight materials having an analogous structure to the polymer, can be used for the estimation of interfacial interaction.

In flow microcalorimetry a small sample is put into the cell of the calorimeter and the probe molecule passes through it in an appropriate solvent. Adsorption of the probe results in an increase in temperature and integration of the area under the signal gives the heat of adsorption [70]. This quantity can be used for the calculation of the reversible work of adhesion according to Eq. 13. The capabilities of the technique can be further increased if a HPLC detector is attached to the microcalorimeter, also the molar heat of adsorption can be determined with this setup.

Lately the most frequently used technique for the determination of thermodynamic and acid/base characteristics is inverse gas chromatography [30,73–75]. In IGC the unknown filler or fiber surface is characterized by compounds, usually solvents, of known properties. IGC measurements can be carried out in two different ways. In the most often applied linear, or ideal, IGC infinite concentrations of n-alkane are injected into the column containing the filler to be characterized. The net retention volume (V_N) can be calculated by:

$$V_N = (t_r - t_0)\ F j_0 \tag{19}$$

where t_r is the retention and t_0 the reference time, F the flow rate of the carrier gas and j_0 is a correction factor taking into account the pressure difference between the two ends of the column. The dispersion component of the surface tension of the filler can be calculated from the retention volumes of n-alkanes:

$$-RT \ln V_n = Na(\gamma_{LV}\ \gamma_s^d)^{1/2} \tag{20}$$

where V_n is the net retention volume of the alkane, a is the surface area of the adsorbed molecule, γ_{LV} the surface tension of the solvent and N is the Avogadro number. The product of RT and the logarithm of the retention volume of normal alkanes is a linear function of $a(\gamma_{LV})^{1/2}$. If the measurements are carried out with polar solvents, the deviation from this straight line is proportional to the acid/base interaction potential of the solid surface [30]. Adsorption isotherms of the probe compound on the filler surface can be determined with non-linear or finite dilution IGC.

When the surface of a solid is only partially wetted by a liquid, it forms a droplet with a definite contact angle (θ). The interaction of the components is expressed by the Young equation:

$$\gamma_{SV} = \gamma_{SL} + \gamma_{LV} \cos\theta \tag{21}$$

where γ_{SV} is the surface tension of the solid in contact with the vapor of the liquid, γ_{LV} is the surface tension of the liquid and γ_{SL} the interfacial tension. Due to wetting, γ_{SV} is smaller than the surface tension of the solid measured in vacuum (γ_{S0}), the difference is the spreading pressure (π_e), i.e.:

$$\gamma_{S0} - \gamma_{SV} = \pi_e \tag{22}$$

The value of π_e is very small for low energy surfaces, but it cannot be neglected for fillers, on the contrary π_e can be used for the calculation of the thermodynamic characteristics of their surface. The spreading pressure can be determined from the adsorption isotherm in the following way [74]:

$$\pi_e = RT \int_0^p \Gamma d \ln p \tag{23}$$

where p is vapor pressure and Γ the moles of vapor adsorbed on a unit volume of the filler. If the measurement is carried out with apolar solvents the dispersion component of the surface tension of the filler can be determined from the spreading pressure:

$$\pi_e = 2(\gamma_s^d \gamma_{LV})^{1/2} - 2\gamma_{LV} \tag{24}$$

Equation 24 is derived from the Young (Eq. 21), the Dupre (Eq. 8) and the Fowkes (Eq. 10) equations by assuming complete wetting (cos θ =0). Measurements with polar solvents give the polar component of the surface tension, but acid/base constants and the corresponding work of adhesion can also be calculated from them:

$$W_{AB}^{spec} = 2\gamma_{LV,p} + \pi_{e,p} - 2(\gamma_s^d \gamma_{LV,p}^d)^{1/2} \tag{25}$$

where W_{AB}^{spec} is the polar or acid/base component of W_{AB} and the p subscript indicates a polar solvent. Finite dilution IGC is more tedious to carry out than linear IGC, but makes possible the direct determination of γ_s^p or the corresponding acid/base constants.

In order to calculate polymer/filler interaction, or more exactly the reversible work of adhesion characterizing it, the surface tension of the polymer must also be known. This quantity is usually determined by contact angle measurements or occasionally the pendant drop method is used. The former method is based on the Young, Dupre and Fowkes equations (Eqs. 21, 8, and 10), but the result is influenced by the surface quality of the substrate. Moreover, the surface (structure, orientation, density) of polymers usually differs from the bulk, which might bias the results. Accuracy of the technique may be increased by using two or more liquids for the measurements. The use of the pendant drop method is limited due to technical problems (long time to reach equilibrium, stability of the polymer, evaluation problems etc.). Occasionally IGC is also used for the characterization of polymers [30].

5.3 Estimation of Interaction

The interaction of two substrates, the bond strength of adhesives are frequently measured by the peel test [76]. The results can often be related to the reversible work of adhesion. Due to its physical nature such a measurement is impossible to carry out for particulate filled polymers. Even interfacial shear strength widely applied for the characterization of matrix/fiber adhesion cannot be used in particulate filled polymers. Interfacial adhesion of the components is usually deduced indirectly from the mechanical properties of composites with the help of models describing composition dependence. Such models must also take into account interfacial interactions.

The number of such models is limited. One of the existing models was developed for the description of the composition dependence of the tensile yield stress of particulate filled composites [5] (see Eq. 17); with some modification the dependence of tensile strength of composites can also be described with the model. Parameter B of the model is related to stress transfer, i.e. to interfacial adhesion. Since interaction depends on the size of the interface (contact area) and the strength of the interaction, both influence the value of B. Parameter B can be determined from the linear plot of the relative yield stress [$\sigma_{yrel}=\sigma_y(1+2.5\varphi_f)/\sigma_{y0}(1-\varphi_f)$] against filler content. The data from Fig. 8 are presented in this form in Fig. 10, the slope of the straight lines gives parameter B and is proportional to interfacial interaction, to the contact area between the filler and the matrix, in this case. Changes in the strength of the interaction can also be expressed by B, as will be shown in Sects. 6.2 and 6.3.

Recently, stress analysis has been carried out for the determination of stress distribution around inclusions in particulate filled composites. A model based on the energy analysis has led to the determination of debonding stress [8]. This stress, which is necessary for the separation of the matrix and filler, was shown to depend on the reversible work of adhesion (see Eq. 16) and it is closely related to parameter B.

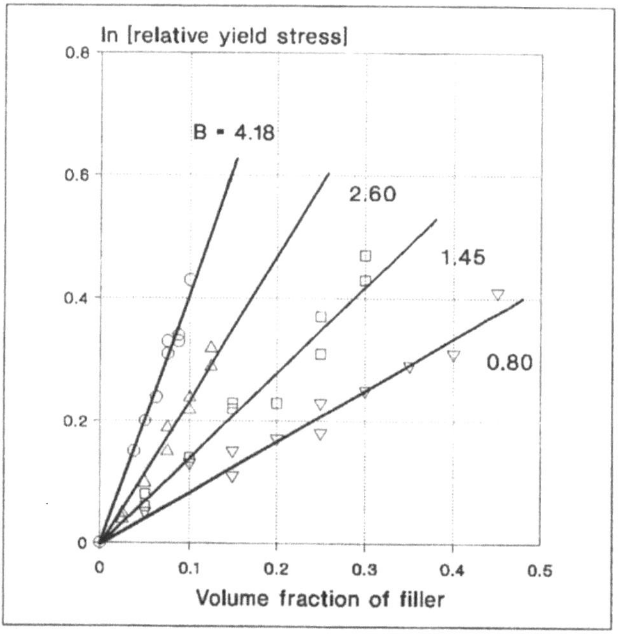

Fig. 10. Quantitative estimation of interaction, effect of the size of the interface. Symbols are the same as in Fig. 8

6 Modification of Interfacial Interactions

The easiest way for the modification of interfacial interactions is the surface treatment of fillers. The compound used for the treatment (coupling agent, surfactant etc.) must be selected according to the characteristics of the components and the goal of the modification. This latter is very important, surface modification is often regarded as a magical tool which can solve all problems of processing technology and product quality. It must always be kept in mind that in a particulate filled polymer two kinds of interaction take place: particle/particle and matrix/filler interaction. Surface treatment modifies both and properties of the composite are determined by the combined effect of the two. As well as the type of material, the amount of the material used for the treatment must also always be optimized both from the technical and the economical aspect. In the following sections surface modifications will be divided into four arbitrary categories and will be discussed accordingly.

6.1 Non-Reactive Treatment – Surfactants

The oldest and most often used modification of fillers is the coverage of their surface with a small molecular weight organic compound [59,77]. Usually amphoteric surfactants are used which have one or more polar groups and a long aliphatic chain. A typical example is the surface treatment of $CaCO_3$ with stearic acid [59,77–79]. The principle of the treatment is the preferential adsorption of the polar group of the surfactant onto the surface of the filler. The high energy surfaces of inorganic fillers (see Table 2) can often enter into special interaction with the polar group of the surfactant. Preferential adsorption is promoted in a large extent by the formation of ionic bonds between stearic acid and the surface of $CaCO_3$, but in other cases hydrogen or even covalent bonds can also form. Surfactants diffuse to the surface of the filler even from the polymer melt, which is further proof for the preferential adsorption [80]. Because of their polarity reactive coupling agents also adsorb onto the surface of the filler. If lack of reactive groups does not make possible chemical coupling with the polymer, these exert the same effect on composite properties as their non-reactive counterparts [56,81–82].

One of the crucial questions of non-reactive treatment, which, however, is very often neglected, is the amount of surfactant to use. It depends on the type of the interaction, the size of the treating molecule, its alignment to the surface and on some other factors. Determination of the optimum amount of surfactant is essential for the efficiency of the treatment. Insufficient amount of surfactant does not bring about the desired effect, while excessive amounts lead to processing problems as well as to the deterioration of the mechanical properties and appearance of the product [74,83].

The amount of bonded surfactant can be determined by simple techniques. A dissolution technique proved to be very convenient for the optimization of non-reactive surface treatment and also for the characterization of the efficiency of the treating technology [74,84]. First the surface of the filler is covered with increasing amounts of surfactant, then the non-bonded part is dissolved with a solvent. The technique is demonstrated in Fig. 11, which presents a dissolution curve of stearic acid on a $CaCO_3$ filler. Surface treatment is preferably carried out with the proportionally bonded surfactant (c_{100}); at this composition the total amount of surfactant used for the treatment is bonded to the filler surface. The filler can adsorb more surfactant (c_{max}), but during compounding a part of it can be removed from the surface by dissolution or simply by shear and might deteriorate properties.

The specific surface area of the filler is an important factor which must be taken into consideration during surface treatment. The proportionally bonded surfactant depends linearly on it [74]. ESCA studies carried out on the surface of a $CaCO_3$ filler covered with stearic acid have shown that ionic bonds form between the surfactant molecules and the filler surface and that the stearic acid molecules are oriented vertically to the surface [74]. These experiments have demonstrated the importance of both the type of the interaction and the alignment of sur-

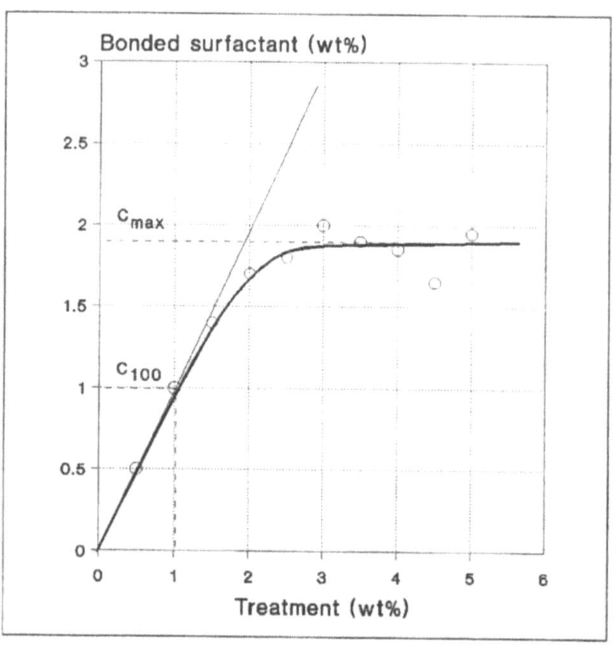

Fig. 11. Typical dissolution curve of stearic acid adsorption on the surface of a $CaCO_3$ filler; r=1.8 μm

factant molecules to the surface. A further proof for the specific character of surface treatment is supplied by the fact that talc and silica adsorb significantly smaller amounts of stearic acid at a unit surface than $CaCO_3$. The lack of specific interaction in the form of ionic bond formation results in a significantly lower amount of proportionally bonded molecules [74].

As a result of the treatment the surface free energy of the filler decreases drastically [59,84]. Lower surface tension also means decreased interfacial tension and reversible work of adhesion, as demonstrated by Fig. 12 where for $PP/CaCO_3$ composites the change of W_{AB} is plotted against the surface coverage of the filler with stearic acid [74]. Such changes in the thermodynamic quantities characterizing interaction result in a decrease of both particle/particle and matrix/particle interaction. One of the main goals, the major reason and benefit of non-reactive surface treatment is the first effect, i.e. to change interaction between the particles of fillers and reinforcements. A weaker interaction leads to a drastic decrease in aggregation, improved dispersion and homogeneity, easier processing, better mechanical properties and appearance. Improvement in the mechanical properties, and first of all in impact strength, is often falsely interpreted as the result of improved wetting and interaction of the components. However, wettability (S) defined as:

$$S_{AB}=\gamma_A-\gamma_B-\gamma_{AB} \tag{26}$$

where subscript A relates to the wetted surface and $\gamma_A > \gamma_B$, decreases with increasing surface coverage of the filler, demonstrated also by the increasing contact angle of organic liquids on the surface of treated fillers [84].

As an effect of non-reactive treatment particle/particle interaction as well as matrix/filler interaction decreases. The consequence of this change is decreased yield stress and strength as well as improved deformability [59,78]. This is demonstrated by Fig. 12 which shows the decrease of tensile strength of $PP/CaCO_3$ composites with increasing surface coverage of the filler. Adhesion and strong interaction, however, are not always necessary or advantageous to prepare composites of desired properties; plastic deformation of the matrix is the main energy absorbing process in impact, which decreases with increasing adhesion [85–87].

6.2 Reactive Treatment – Coupling Agents

Reactive surface treatment assumes chemical reaction of the coupling agent with both of the components. The considerable success of silanes in glass reinforced thermosets have led to their application in other fields; they are used, or at least experimented with, in all kinds of composites irrespective of the type, chemical composition or other characteristics of the components. Reactive treatment, however, is even more complicated than non-reactive; polymerization of the coupling agent, development of chemically bonded and physisorbed layers render the identification of surface chemistry, characterization of the interlayer

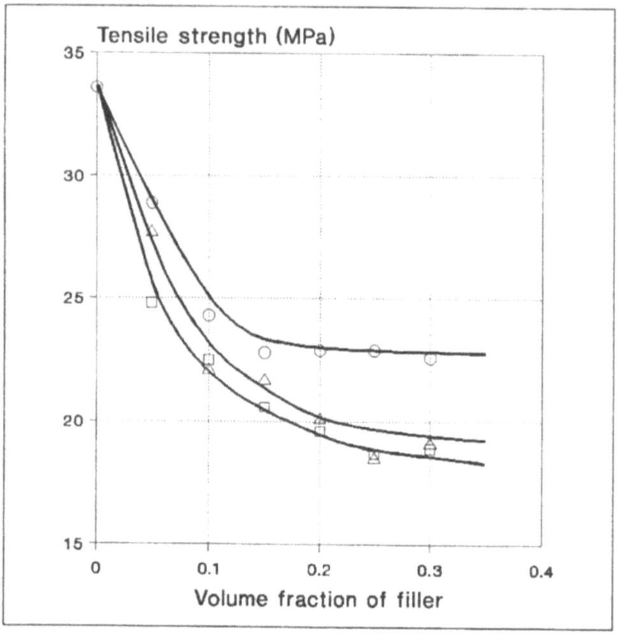

Fig. 12. Effect of non-reactive surface treatment of a $CaCO_3$ filler with stearic acid on its interaction with PP, r=1.8 μm; (○) non-treated, (△) 75%, (□) 100% surface coverage

and optimization of the treatment very difficult [65]. In spite of these difficulties significant information has been collected on silanes [88]; much less is known about the mechanism and effect of other coupling agents such as titanates, zirconates, etc. [77].

Silane coupling agents are successfully used with fillers and reinforcements which have reactive OH groups on their surface, e.g. glass fibers, glass flakes and beads, mica and other silicate fillers [43,89,90]. Use of silanes with fillers like $CaCO_3$, $Mg(OH)_2$, wood flour, etc. were tried, but proved to be unsuccessful in most cases [91,92]; sometimes contradictory results were obtained even with glass and also other siliceous fillers [93]. Acidic groups are preferable for $CaCO_3$, $Mg(OH)_2$, $Al(OH)_3$ and $BaSO_4$. Talc cannot be treated successfully either with reactive or non-reactive agents because of its inactive surface, only broken surfaces contain a few active OH groups. The chemistry of silane modification has been extensively studied and described elsewhere [65,88]. Model experiments have shown that a multilayer film forms on the surface of the filler; the first layer is chemically coupled to the surface, this is covered by cross-linked silane polymer, and the outer layer is physisorbed silane. The matrix polymer may react chemically with the coupling agent, but interdiffusion of the matrix and the polymerized silane coupling agent, i.e. physical interaction, also takes place [94,95].

Recent studies have shown that the adsorption of organofunctional silanes is usually accompanied by polycondensation. The high energy surface of non-treated mineral fillers bonds water from the atmosphere, which initiates the hydrolysis and polymerization of the coupling agent [96,97]. FTIR and DRIFT spectra of a $CaCO_3$ filler treated with an aminosilane coupling agent is presented in Fig. 13. All four spectra are the same, in spite of different treatment and sample preparation conditions. Spectrum A was recorded on a filler treated from n-butanol solution, while Specta B and C were taken from a filler treated by dry-blending. The treated filler was washed with THF and the similarity of the two spectra (B and C) shows that the polysiloxane layer adsorbs strongly to the surface. Moreover, this spectrum is the same as the one recorded on a polymerized film of the neat aminosilane. The experiments prove the fact of the polycondensation, and also that strong coupling to the filler can be achieved even if the filler does not contain active OH groups on its surface. However, the adsorbed amount of coupling agent, structure, properties and thus adhesion of the polysiloxane layer depend very much on the chemical composition of the organofunctional group of the coupling agent. This is obvious if we compare the spectra presented in Figs. 9 and 13, and also if we study the dissolution curve of the two silanes on $CaCO_3$ (Fig. 14). The different chemical structures of the silanes lead to

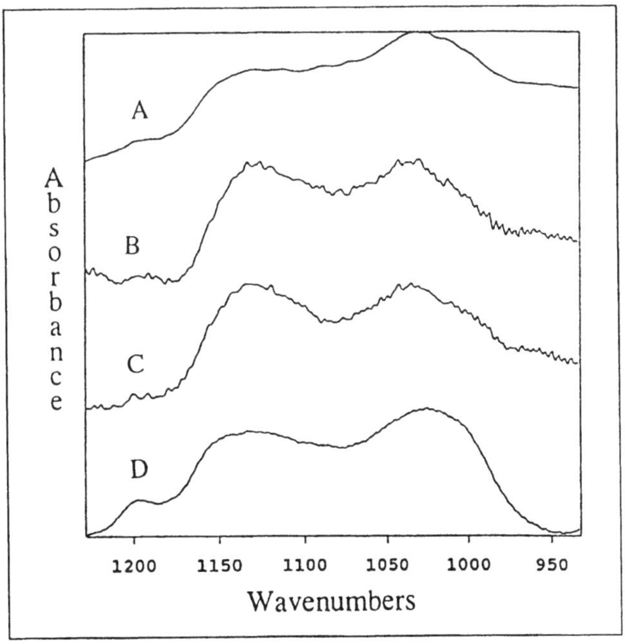

Fig. 13. A) DRIFT difference spectrum of $CaCO_3$ treated with 0.8 wt% AMPTES form n-butanol solution; FTIR difference spectra of $CaCO_3$ treated with 1 wt% AMPTES B) before washing with THF, C) after washing with THF, D) spectrum of condensed neat AMPTES

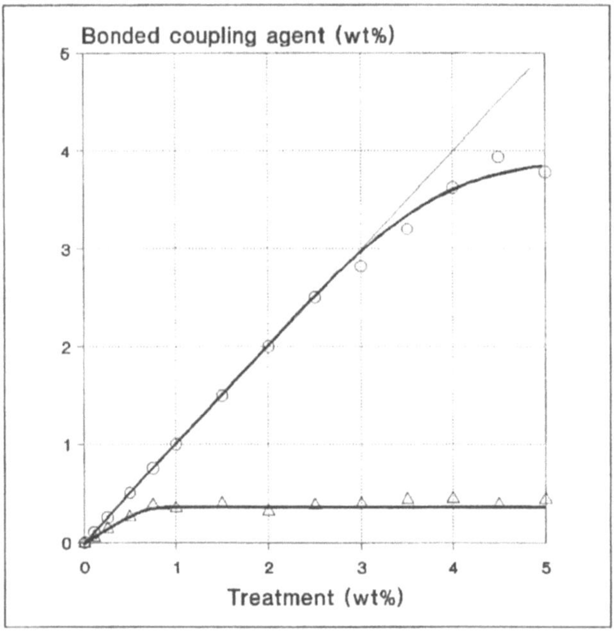

Fig. 14. Effect of the chemical structure of silane coupling agents on their adsorption on the surface of a $CaCO_3$ filler, r=1.25 μm; (○) aminosilane (AMPTES), (△) aliphatic silane (SPTES)

considerably different adsorptions. The figure shows also that the simple dissolution technique can be advantageously applied for the study of reactive coupling agents as well [96].

Although the chemistry of silane modification of reactive silica fillers is well documented, much less is known about the interaction of silanes with polymers. It is more difficult to create covalent bonds between a coupling agent and a thermoplastic, since the latter rarely contains reactive groups. The most reactive are the polycondensation polymers, i.e. polyesters or polyamides, where transesterification reactions may lead to coupling. Literature references indicate some evidence of reactive coupling; indeed, the strength of polyamide and polycarbonate composites increases on aminosilane treatment [97–99]. The increased strength is always interpreted as an improvement of properties. However, the change in deformability and impact characteristics are usually not mentioned; they frequently decrease as an effect of reactive treatment as has also been shown [99].

Reactive treatment is the most difficult in polyolefins, although occasionally successful attempts have been reported as shown in Fig. 15 for a PP/mica composite [89]; the results presented in the figure were obtained by the application of an aminosilane coupling agent. In PP composites the most often used cou-

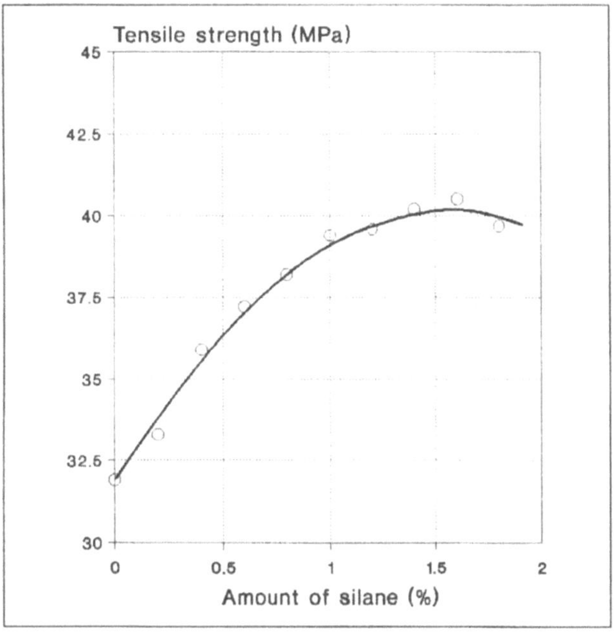

Fig. 15. Reactive coupling in PP/mica composites [89]

pling agent is the Z6032 product from Dow Corning, which is N-β-(N-vinylbenzylamino)ethyl-γ-aminopropyltrimethoxysilane [43,91]. Although one would expect that coupling occurs through the reaction of free radicals as suggested by Widman et al. [99], the latest results show that polypropylene oxidizes during processing even in the presence of stabilizers and the formed acidic groups react with aminosilanes leading to reactive coupling [100]. The presence of reactive coupling and the strength of the interaction can be proved by the application of the model presented in Sect. 5.3. Figure 16 shows the linear plot of relative yield stress against the volume fraction of the filler for three different treatments in PP/$CaCO_3$ composites. It is obvious that stearic acid acts as a surfactant, while the aminosilane is a reactive coupling agent. The strength of the interaction is reflected well by parameter B.

Considering the complexity of the chemistry involved, it is not surprising that the amount of coupling agent and surface coverage have an optimum here too, similar to the surfactants in non-reactive surface treatment. This has been proved by Trotignon and Verdu [89] who found a strong increase and a maximum in the strength of PP/mica composites as a function of the silane used (see Fig. 15). Optimization of the type and amount of coupling agent is also crucial in reactive treatment and although "proprietary" treatments may lead to some improvement in properties, they might not be optimal or cost effective. The im-

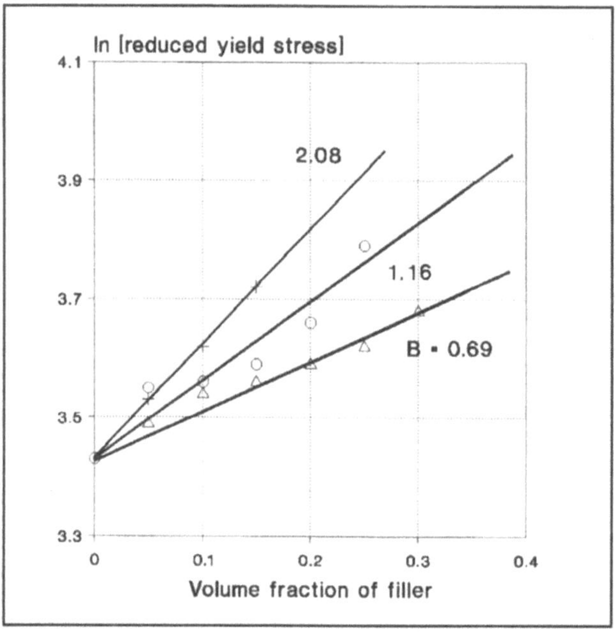

Fig. 16. Effect of treatment on the tensile yield stress of PP/$CaCO_3$ composites. Treatment: (△) stearic acid, (○) no treatment, (+) aminosilane

proper choice of coupling agent may lead to insufficient or even detrimental effects. In some cases hardly any change in the properties are observed or the effect can be clearly attributed to the decrease of surface tension due to the coverage of the filler surface by an organic substance, i.e. non-reactive treatment [81,82].

6.3 Polymer Layer – Interdiffusion

Numerous results indicate that physical adsorption of the polymer on the filler surface leads to the development of a rigid interphase (see Sect. 4). Chemical coupling with a single covalent bond has the same effect. The coverage of filler surface with a polymer layer which is capable of interdiffusion with the matrix proved to be very effective both in stress transfer and in forming a thick, diffuse interphase with acceptable deformability. In this treatment the filler is usually covered by a functionalized polymer, preferably by the same polymer as the matrix. The functionalized polymer is attached to the surface of the filler by secondary, hydrogen, ionic and sometimes by covalent bonds. The polymer layer interdiffuses with the matrix, entanglements are formed and strong adhesion is created. Because of its increased polarity, in some cases reactivity, maleic anhydride or acrylic acid modified polymers are used, which adsorb to the surface of most polar fillers even from the melt. Most frequently this treatment is used in polyolefin composites, since, on the one hand, other treatments often fail, and, on the other, functionalization of these polymers is relatively easy. Often a very small amount of modified polymer is sufficient to achieve significant improvement in stress transfer [101,102].

The importance of interphase interdiffusion and entanglement density is clearly demonstrated by the experiments of Felix and Gatenholm [103], who introduced maleic anhydride modified PP (MA-PP) into PP/cellulose composites and achieved an improvement in yield stress. The increase was proportional to the molecular weight of MA-PP (Fig. 17). Increased adhesion is clearly shown by the increase in parameter B, which reflects the strength of the interaction in this case (Fig. 18). Model calculations have shown that the improvement of properties has an upper limit and a plateau is reached at a certain, not too high molecular weight. Maximum effect of functionalized PP was found with fillers of high energy surfaces [93,104,105] or with those capable of specific interactions, e.g. ionic bond with $CaCO_3$ [102,106] or chemical reaction with wood flour, kraft lignin or cellulose [101,103].

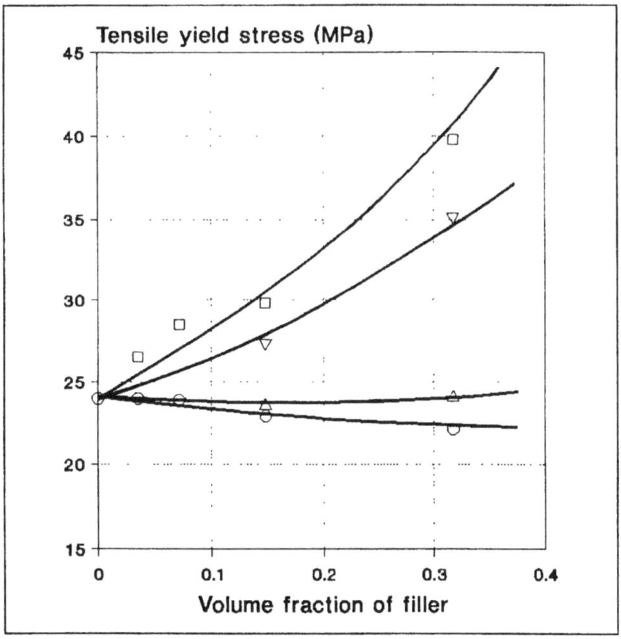

Fig. 17. Effect of interdiffusion of the functionalized polymer with the matrix on the mechanical properties of PP/cellulose composites. Molecular weight of MA-PP: (○) non-treated, (△) 350, (▽) 4500, (□) 3.9×10^4

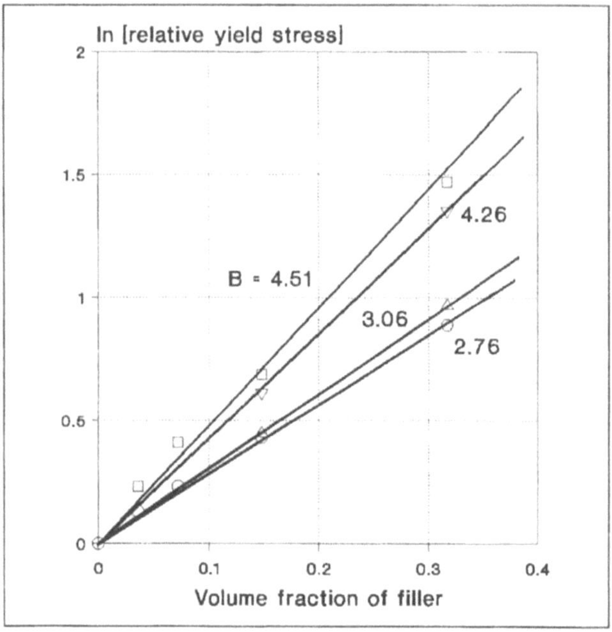

Fig. 18. Increased adhesion due to the coverage of the filler with a functionalized polymer. Symbols are the same as in Fig. 17

6.4 Soft Interlayer – Elastomers

Incorporation of hard particles into the polymer matrix creates stress concentration, which induces local micromechanical deformation processes. Occasionally these might be advantageous for increasing plastic deformation and impact resistance, but usually they cause deterioration in the properties of the composite. Encapsulation of the filler particles by an elastomer layer changes the stress distribution around the particles and modifies the local deformation processes. Encapsulation can take place spontaneously, it can be promoted by the use of functionalized elastomers (see Sect. 6.3) or the filler can be treated in advance.

The coverage of the filler with an elastomer layer has been studied mostly in PP composites, but occasionally in other polymers as well. PP has a poor low temperature impact strength, which is frequently improved by the introduction of elastomers [107]. Improvement in impact strength, however, is accompanied by a simultaneous decrease of modulus, which cannot be accepted in certain applications; a filler or reinforcement is added to compensate the effect. Although most of the papers dealing with these materials agree that the simultaneous introduction of the two different types of material (elastomer, filler) is beneficial, this is practically the only similarity which can be found in them. A large number of such systems were prepared and investigated, but the observations concerning their structure, the distribution of the components and their effect on composite properties are rather controversial. In some cases separate distribution of the components and independent effects were observed [108,109], in others encapsulation of the filler by the elastomer [110,111]. In spite of its importance, structure of the composites is not always clear [112,113]. These different structures naturally lead to dissimilar properties as well. Composition dependence of shear modulus agrees well with values predicted by the Lewis-Nielsen model in the case of separate dispersion of the components, while large deviations are observed when filler particles are embedded into the elastomer [114]. Embedding of the filler particles was observed also in the dynamic mechanical spectra of the composites, encapsulated filler particles extended the elastomer leading to an increased relaxation peak maximum at around −50 °C. Model calculations have shown that a maximum of 70% of the filler particles could be embedded into the elastomer, while even in the cases when separate dispersion of the components dominates, at least 5–10% of the particles were encapsulated. The results indicate that complete encapsulation or separate dispersion cannot be achieved by the mere combination and homogenization of the components.

Extensive investigations [109,115,116] have shown that the final structure is determined by the relative magnitude of adhesion and shear forces during the homogenization of the composite. Similar principles can be used as for the estimation of aggregation, since the same forces are effective in the system. If shear forces are larger than adhesion, separate distribution of the components occurs, while strong adhesion leads to embedded structure (see Eq. 7). The three parameters which influence structure are adhesion (W_{AB}), particle size (r) and shear rate ($\dot{\gamma}$). Fillers usually have a wide particle size distribution, very often in the 0.3–10 μm range. As a consequence, composites will always contain small particles which are encapsulated and large particles which are separately distributed [115].

All the experiments carried out on such systems have shown the primary importance of adhesion in structure formation and in the resulting effect on properties. Numerous attempts have been made to prepare composites with exclusive structure, i.e. complete coverage or separate distribution. In most cases PP or elastomers modified with maleic anhydride or acrylic acid were used to enhance adhesion between selected components; separate dispersion in the case of MA-PP, embedding with MA-EPDM [104,105,116,117]. Although the desired effect was observed in almost every case, the exclusiveness could not be proved. Similarly contradictory is the effect on properties, especially on impact resistance. In three-component PE composites a better impact resistance/modulus ratio could be achieved, but Kelnar [106] observed lower impact strength with AA-EPDM than with a non-modified elastomer. Although the relationship of structure and impact resistance has not yet been determined unambiguously, experiments carried out on a series of PP bumper compounds containing a filler and an elastomer have proved that simultaneous increase of impact resistance and stiffness

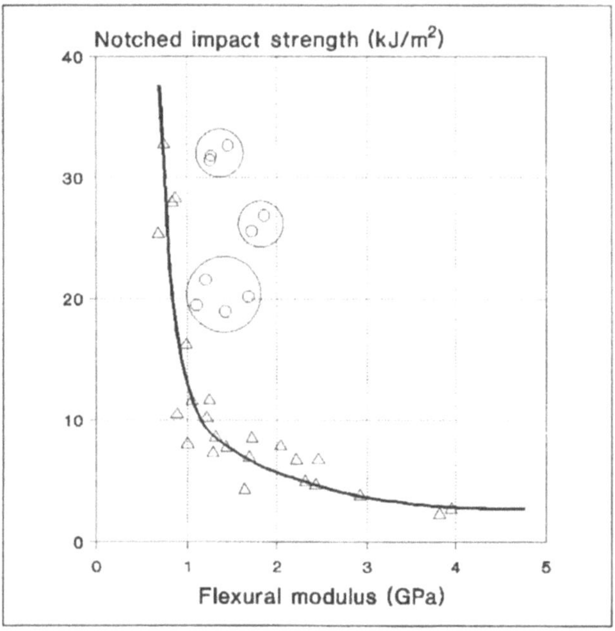

Fig. 19. Improved stiffness and impact resistance in PP composites containing a filler and an elastomer; (△) usual behavior, (○) improved properties

can be achieved by the introduction of a separate elastomer phase (Fig. 19). Such composites show a positive deviation from the usual, inverse correlation of stiffness and impact resistance.

7 Practical Relevance

Interactions taking place in particulate filled composites strongly influence their properties. Interactions are responsible for aggregation and the formed interphase has a strong effect on composite performance. Interphase, or more exactly the immobilized polymer layer, has the same effect as increased filler content. This increased apparent or effective filler content leads to the decrease in the maximum amount of filler which can be introduced into the polymer matrix [59]. Taking into account the spatial arrangement of the particles the maximum filling grade, i.e. maximum packing fraction (φ_f^{max}) can be calculated. In the case of spherical particles it ranges from 0.52 to 0.74, while it can go up as high as 0.91 for anisotropic fillers and reinforcements [17,118]. In practice, however, incorporation of even much less filler leads to insurmountable difficulties or useless material. Figure 20 shows the maximum packing fraction of some $CaCO_3$

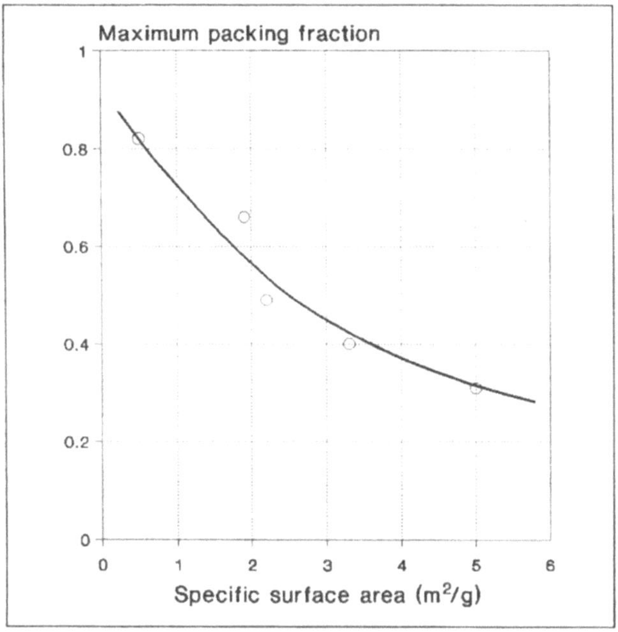

Fig. 20. Dependence of the maximum packing fraction on the specific surface area of the filler in PP/$CaCO_3$ composites

fillers calculated from experimental tensile yield stress data [59] as a function of the specific surface area of the filler. It is obvious that the value of φ_f^{max} does not depend on packing since all the particles have practically the same, more or less spherical, geometry and similar relative size distribution. Even the effect of aggregation can be excluded [17], since these fillers have relatively low specific surface area and do not show any tendency for aggregation.

The effect of the size of the interface, i.e. specific surface area of the filler, on the properties of the composite offer a possibility to prepare tailor-made composites with desired properties. The adjustment of composite properties in this way, however, has limitations both in the lower and the upper range of specific surface area. The adhesion of large filler particles (small specific surface area) to the polymer matrix is weak, debonding takes place even under the effect of moderate load [7,8]. Moreover, cavities formed around large particles merge to critical cracks very fast, which leads to decreased strength and premature failure [6]. Reinforcing thermoplastics with spherical fillers would be very attractive, but in order to achieve large specific surface area, the particle size of the filler must be small. Small particles, however, form aggregates creating crack initiation sites and increased notch sensitivity [6,64,119]. Due to these effects, the specific surface area range for non-treated spherical fillers, which can be utilized in PP, is rather narrow, it is somewhere between 2 and 5 m^2/g [64].

Surface treatment is obviously the most convenient route to modify interaction and thus composite properties. A strong thin interphase results in brittle materials, while a soft interphase leads to lower modulus and higher impact resistance. Mostly non-reactive treatment is used in thermoplastics. The cost of reactive treatment is not justified in most cases by the achieved technical advantages, although at high filler loadings in specific applications (flame retarded composites) it might be advantageous [120,121]. The application of functionalized polymers and the introduction of an elastomer interlayer gains more and more importance in all kinds of composites [120,121]. Proper treatment can strongly improve the performance of the composite and facilitate processing.

8 Conclusions

According to some authors interfacial interaction is the determining factor in all heterogeneous polymer systems, including particulate filled polymers. Both particle/particle and matrix/filler interactions play an important role in the determination of composite properties. The effect of the former is clearly detrimental, it decreases strength and, in particular, impact resistance of the composite. The occurrence and extent of aggregation is determined by the relative adhesion and shear forces during the homogenization of the components. Due to matrix/filler interaction an interphase forms spontaneously in composites with properties different from that of both components. The strength of adhesion can be acceptably characterized by thermodynamic quantities, mainly by the reversible work of adhesion. The modification of interactions is achieved through the surface treatment of the filler. In particulate filled systems, non-reactive treatment is applied the most often, but lately other techniques have also gained importance. The type of surface treatment must be selected according to the goal, the surfactant or coupling agent must be adjusted according to the chemical character of the components and the optimum amount of treatment must be determined in each case. Proper modification may improve properties significantly, occasionally it can be the condition of the successful application of the composite.

9 References

1. Vink D (1990) Kunststoffe 80: 842
2. Pukánszky B (1995) In: Karger-Kocsis J (ed) Polypropylene. Structure, Blends and Composites, vol 3. Chapman and Hall, London, p 1
3. Krysztafkiewicz A (1988) Surface Coatings Technol 35: 151
4. Vörös G, Fekete E, Pukánszky B (1997) J. Adhesion 64: 229
5. Pukánszky B, Turcsányi B, Tüdős F (1988) In: Ishida H (ed) Interfaces in polymer, ceramic, and metal matrix composites. Elsevier, New York, p 467
6. Pukánszky B (1990) Composites 21: 255
7. Vollenberg P, Heikens D, Ladan HCB (1988) Polym Compos 9: 382

8. Pukánszky B, Vörös G (1993) Compos Interfaces 1: 411
9. Riley AM, Paynter CD, McGenity PM, Adams JM (1990) Plast Rubber Process Appl 14: 85
10. Menczel J, Varga J (1983) J Thermal Anal 28: 161
11. Fujiyama M, Wakino T (1991) J Appl Polym Sci 42: 2739
12. Pukánszky B, Mudra I, Staniek P (1997) J Vinyl Additive Technol 3: 53
13. Bezerédi Á, Mudra I, Pukánszky B, Structure/Property Correlations in Nucleated PP Polymers; Impact Resistance, submitted to J Appl Polym Sci
14. Schlumpf HP (1990) Chimia 44: 359
15. Rothon R (1995) Particulate-filled polymer composites. Longman, Harlow
16. Schlumpf HP (1983) Kunststoffe 73: 511
17. Nielsen LE (1974) Mechanical properties of polymers and composites. Marcel Dekker, New York
18. Svehlova V, Poloucek E (1987) Angew Makromol Chem 153: 197
19. Ess JW, Hornsby PR (1987) Plast Rubber Process Appl 8: 147
20. Adams MJ, Mullier MA, Seville JPK (1987) In: Briscoe BJ, Adams MJ (eds) Tribology in particulate technology. Adam Hilger, Bristol, p 375
21. Tabor D (1987) In: Briscoe BJ, Adams MJ (eds) Tribology in particulate technology. Adam Hilger, Bristol, p 206
22. Adams MJ, Edmondson B (1987) In: Briscoe BJ, Adams MJ (eds) Tribology in particulate technology. Adam Hilger, Bristol, p 154
23. Balachandran W (1987) In: Briscoe BJ, Adams MJ (eds) Tribology in particulate technology. Adam Hilger, Bristol, p 135
24. Goren SL (1971) J Colloid Interface Sci 36: 94
25. Pukánszky B, Fekete E (1997) Aggregation tendency of particulate fillers: determination and consequences, paper presented at Eurofillers 1997, September 7–11, Manchester
26. Allen KW (1988) Phys Technol 19: 234
27. Wu S (1978) In: Paul DR, Newman S (eds) Polymer blends. Academic Press, New York, vol 1, p 243
28. Allen KW (1987) J Adhesion 21: 261
29. Derjaugin BV (1955) Research 8: 70, 365
30. Schreiber HP (1993) In: Akovali G (ed) The interfacial interfacial interactions in polymeric composites. Kluwer, Amsterdam, p 21
31. Good RJ (1977) J Colloid Interface Sci 59: 398
32. Fowkes FM (1964) Ind Eng Chem 56: 40
33. Fowkes FM (1968) In: Fowkes FM (ed) Hydrophobic surfaces. Proc Kendall Award Symp, p 151
34. Wu S (1974) J Macromol Sci, Rev Macromol Chem C6: 85
35. Mittal KL, Anderson HR (1991) Acid-base interactions: relevance to adhesion science and technology. VSP, Utrecht
36. Fowkes FM (1981) In: Mittal KL (ed) Physicochemical aspects of polymer surfaces. Plenum, New York, p 583
37. Fowkes FM (1991) In: Mittal KL, Anderson HR (eds) Acid-base interactions: relevance to adhesion science and technology. VSP, Utrecht, p 93
38. Drago RS, Vogel GC, Needham TE (1971) J Am Chem Soc 93: 6014
39. Gutmann V (1978) The donor-acceptor approach to molecular interactions. Plenum, New York
40. Mittal KL (1977) Polym Eng Sci 17: 467
41. Pukánszky B, Vörös G (1996) Polym Compos 17: 384
42. Fox HW, Hare EF, Zismann WA (1955) J Phys Chem 59: 1097
43. Yue CY, Cheung WL (1991) J Mater Sci 26: 870
44. Vollenberg PHT, Heikens D (1989) Polymer 30: 1656
45. Stamhuis JE, Loppé JPA (1982) Rheol Acta 21: 103
46. Sumita M, Tsukini H, Miyasaka K, Ishikawa K (1984) J Appl Polym Sci 29: 1523

47. Zorll U (1977) Gummi, Asbest, Kunstst 30: 436
48. Folkes MJ, Wong WK (1987) Polymer 28: 1309
49. Akay G (1990) Polym Eng Sci 30: 1361
50. Maurer FHJ, Kosfeld R, Uhlenbroich T (1985) Colloid Polym Sci 263: 624
51. Maurer FHJ, Kosfeld R, Uhlenbroich T, Bosveliev LG (1981) 27th Intl. Symp. on Macromolecules, 6–9 July, Strasbourg, France
52. Mansfield KF, Theodorou DN (1991) Macromolecules 24: 4295
53. Patel S, Hadziioannou G, Tirrell M (1986) In: Ishida H, Koenig JL (eds) Composite interfaces. Elsevier, New York, p 65
54. Jančář J (1991) J Mater Sci 26: 4123
55. Vollenberg PHT, Heikens D (1986) In: Ishida H, Koenig JL (eds) Composite interfaces. Elsevier, New York, p 171
56. Maiti SN, Mahapatro PK (1991) J Appl Polym Sci 42: 3101
57. Vörös G, Pukánszky B (1995) J Mater Sci 30: 4171
58. Maurer FHJ, Schoffeleers HM, Kosfeld R, Uhlenbroich T (1982) In: Hayashi T, Kawata K, Umekawa S (eds) Progress in science and engineering of composites. ICCM-IV, Tokyo, p 803
59. Pukánszky B, Fekete E, Tüdõs F (1989) Makromol Chem, Macromol Symp 28: 165
60. Iisaka K, Shibayama K (1978) J Appl Polym Sci 22: 3135
61. Pukánszky B, Tüdös F (1990) In: Ishida H (ed) Controlled interphases in composite materials. Elsevier, New York, p 691
62. Kolařík J, Hudeček S, Lednický F (1978) Faserforsch Textiltechn 29: 51
63. Kolařík J, Hudeček S, Lednický F, Nicolais L (1979) J Appl Polym Sci 23: 1553
64. Pukánszky B (1992) New Polym Mater 3: 205
65. Ishida H, Koenig JL (1980) J Polym Sci, Polym Phys 18: 1931
66. Briggs D, Seah MP (1990) Practical surface analysis, vol 1, Auger and X-ray photoelectron spectroscopy. Wiley, Chichester
67. Briggs D, Seah MP (1992) Practical surface analysis, vol 2, Ion and neutral spectrocopy. Wiley, Chichester
68. Banwell CN (1972) Fundamentals of molecular spectroscopy. McGraw Hill, New York
69. Briggs D, Brown A, Vickerman JC (1989) Handbook of static secondary ion mass spectrometry. Wiley, Chichester
70. Ashton DP, Briggs D (1995) in Ref 15, p 89
71. Eltekova NA (1995) Proc Eurofillers '91, p 285
72. Delon JF, Yvon J, Michot L, Villieras F, Dases JM (1995) Proc Eurofillers '95, p 17
73. Balard H, Papirer E (1993) Progr Surface Coatings 22: 1
74. Fekete E, Pukánszky B, Tóth A, Bertóti I (1990) J Colloid Interface Sci 135: 200
75. Panzer U, Schreiber HP (1992) Macromolecules 25: 3633
76. Bonnerup C, Gatenholm P (1993) J Adhesion Sci Technol 7: 247
77. Rothon RN (1995) in Ref 15, p 123
78. Jančář J, Kučera J (1990) Polym Eng Sci 30: 707
79. Jančář J (1989) J Mater Sci 24: 3947
80. Marosi G, Bertalan G, Rusznák I, Anna P (1986) Colloids Surf 23: 185
81. Bajaj P, Jha NK, Jha RK (1989) Polym Eng Sci 29: 557
82. Bajaj P, Jha NK, Jha RK (1989) Br Polym J 21: 345
83. Raj RG, Kokta BV, Dembele F, Sanschagrin B (1989) J Appl Polym Sci 38: 1987
84. Papirer E, Schultz J, Turchi C (1984) Eur Polym J 12: 1155
85. Allard RC, Vu-Khanh T, Chalifoux JP (1989) Polym Compos 10: 62
86. Bramuzzo M, Savadori A, Bacci D (1985) Polym. Compos. 6: 1
87. Pukánszky B, Maurer FHJ (1995) Polymer 36: 1617
88. Plueddemann EP (1982) Silane coupling agents, Plenum, New York
89. Trotignon JP, Verdu J, De Boissard R, De Vallois A (1986) In: Sedláček B (ed) Polymer Composites. Walter de Gruyter, Berlin, p 191
90. Matienzo LJ, Shah TK (1986) Surf Interface Anal 8: 53

91. Vollenberg PHT, Heikens D (1990) J Mater Sci 25: 3089
92. Raj RG, Kokta BV, Daneault C (1989) Int J Polym Mater 12: 239
93. Mäder E, Freitag KH (1990) Composites 21: 397
94. Ishida H, Miller JD (1984) Macromolecules 17: 1659
95. Ishida H (1985) In: Ishida H, Kumar G (eds) Molecular characterization of composite interfaces. Plenum
96. Demjén Z, Pukánszky B, Földes E, Nagy J (1997) J Interface Colloid Sci 194: 269
97. Sadler EJ, Vecere AC (1995) Plast Rubber Process Appl 24: 271
98. Zolotnitsky M, Steinmetz JR (1995) J Vinyl Additive Technol 1: 109
99. Widmann B, Fritz HG, Oggermüller H (1992) Kunststoffe 82: 1185
100. Demjén Z, Pukánszky B, Nagy JJr (1997) Possible coupling reactions of functional silanes and polypropylene, accepted in Polymer
101. Takase S, Shiraishi N (1989) J Appl Polym Sci 37: 645
102. Jančář J, Kummer M, Kolařík J (1988) In: Ishida H (ed) Interfaces in polymer, ceramic, and metal matrix composites. Elsevier, New York, p 705
103. Felix JM, Gatenholm P (1991) J Appl Polym Sci 50: 699
104. Kelnar I (1991) Angew Makromol Chem 189: 207
105. Chiang WY, Yang WD (1988) J Appl Polym Sci 35: 807
106. Jančář J, Kučera J (1990) Polym Eng Sci 30: 714
107. Karger-Kocsis J, Kalló A, Kuleznev VN (1984) Polymer 25: 279
108. Lee YD, Lu CC (1982) J Chin Inst Chem Eng 13: 1
109. Kolařík J, Lednický F (1986) In: Sedláèek B (ed) Polymer composites. Walter de Gruyter, Berlin, p 537
110. Stamhuis JE (1988) Polym Compos 9: 280
111. Gupta AK, Kumar PK, Ratnam BK (1991) J Appl Polym Sci 42: 2595
112. Serafimov B (1986) Plaste Kautsch 33: 331
113. Faulkner DL (1988) J Appl Polym Sci 36: 467
114. Kolařík J, Lednický F, Pukánszky B (1987) In: Matthews FL, Buskell NCR, Hodgkinson JM, Morton J (eds) Proc 6th ICCM/2nd ECCM. Elsevier, London, vol 1, p 452
115. Pukánszky B, Tüdös F, Kolařík J, Lednický F (1990) Polym Compos 11: 98
116. Chiang WY, Yang WD, Pukánszky B (1992) Polym Eng Sci 32: 641
117. Kolařík J, Lednický F, Jančář J, Pukánszky B (1990) Polym Commun 31: 201
118. Turcsányi B, Pukánszky B, Tüdós F (1988) J Mater Sci Lett 7: 160
119. Michler GH, Tovmasjan JM Plaste Kautsch (1988) 35: 73
120. Liauw CM, Lees GC, Hurst SJ, Rothon RN, Dobson DC (1995) Plast Rubber Compos Process Appl 24: 211
121. Liauw CM, Lees GC, Hurst SJ, Rothon RN, Dobson DC (1995) Plast Rubber Compos Process Appl 24: 249

Received: April 1998

Rheology, Compounding and Processing of Filled Thermoplastics

P.R. Hornsby

Department of Materials Engineering, Brunel University, Uxbridge, UB8 3PH, UK
e-mail: Peter.Hornsby@brunel.ac.uk

The influence of fillers on the rheology of polymer melts is reviewed, together with an account of mechanisms involved during their combination in melt-mixing procedures. The application of these principles to the design and operation of industrial compounding technologies is then discussed. Means for inducing further microstructural changes during secondary melt processing are described, leading to the achievement of enhanced composite performance.

Keywords: rheology, fillers, mixing thermoplastics, structure, processing

1
Scope of Review

Fillers are introduced into thermoplastics for many varied reasons, more generally to influence the physical properties of the polymer, but occasionally to simply act as extenders or matrix diluents.

Whilst the physical and chemical nature of the filler will determine its effectiveness in a functional role, the presence of solid additives in thermoplastics melts inevitably influence their processability. The extent to which this occurs depends on many factors including the amount of filler present, possible interactive effects between the filler and polymer, or between the filler particles themselves, together with the conditions experienced during melt processing, in particular the shear and/or elongational flow fields developed.

The end performance of the filler is also critically influenced by how it is presented within the polymer especially the state of mixing within the composition. With some filler types it is possible to enhance their efficiency, or commercial viability, by the way in which they are combined and melt processed with the polymer. In some instances this has been achieved through innovation in processing machinery design.

This chapter will consider these issues through an appraisal of how fillers can influence the rheological properties of molten polymers, current understanding of fundamental principles governing the mixing of particulate additives and thermoplastics melts, and the implications of this knowledge on the engineering design and effective operation of industrial compounding plant. In addition, techniques will be highlighted which can further change the microstructure within filled polymer compositions, particularly during secondary melt processing operations.

2 Rheology of Filled Polymers

2.1 Introduction

Extensive literature exists on the rheology of suspensions containing rigid fillers, which includes discussion of a wide range of filler types and fluid systems exposed to various flow regimes. Of particular relevance to the present discussion is the influence of a filler on the rheological properties of viscous polymer melts, and the consequences of such behaviour on subsequent mixing and forming operations, to yield products with defined properties. Hence an understanding of the rheology of filled polymers is beneficial in the design of polymer conversion equipment, through an estimation of pressure drops, forces and power requirements, the optimization of processing conditions and correlation with the structure developed in the filled composition.

Since polymer melt flow behaviour is strongly affected by the nature of the filler type, including its morphology, surface chemistry and concentration, rheological studies can also assist in the development of formulations designed to facilitate industrial processability.

Rheological properties of filled polymers can be characterised by the same parameters as any fluid medium, including shear viscosity and its interdependence with applied shear stress and shear rate; elongational viscosity under conditions of uniaxial extension; and real and imaginary components of a complex dynamic modulus which depend on applied frequency [1]. The presence of fillers in viscoelastic polymers is generally considered to reduce melt elasticity and hence influence dependent phenomena such as die swell [2].

The existence of a yield stress is a common feature of highly filled polymer melts, which is associated with particle interactions within the matrix. At stresses below a threshold value the material has unbounded viscosity generally behaving like a solid undergoing only elastic deformation, whereas at higher applied stress levels it can experience flow. The degree of filler dispersion also contributes to the rheological properties of polymer melts containing interactive particles.

Specific difficulties must be addressed for meaningful rheological measurement of filled polymer systems. Firstly, the dimensions of suspended particles must be significantly smaller than the characteristic dimensions of the measuring equipment within which flow occurs. Additionally, inertial and gravitational effects should be negligible. A general phenomenon, typically associated with the hydrodynamics of filled polymers, is the wall effect caused by a nonhomogeneous distribution of the disperse phase resulting in the formation of a melt layer at the wall surface, which is depleted of filler. This layer of relatively low viscosity existing at the melt boundary gives rise to lubrication effects or apparent wall slippage [1].

Further factors influencing rheological characterization of filled polymers include changes in the degree of filler dispersion or inter-particle structure formation during the period of measurement, and interactive effects between the filler phase and surrounding matrix. These may originate, for example, from shear-induced crystallization of the polymer, or immobilised layers of polymer adsorbed onto the filler surface.

It will be apparent from the foregoing discussion that the suspension of filler particles within a fluid polymeric phase requires consideration of issues beyond the usual treatment of rheology applied to non-Newtonian polymer melts.

2.2 Newtonian Suspensions Containing Rigid Fillers

The starting point for discussing the rheology of fluid suspensions containing rigid fillers is generally the Einstein equation [3], which predicts the viscosity of a Newtonian fluid containing a very dilute suspension of rigid spheres:

$$\eta = \eta_1 \left(1 + k_E \phi\right) \tag{1}$$

where η is the viscosity of the mixture, η_1 is the viscosity of the suspending liquid and ϕ is the volume fraction of filler. k_E is the Einstein coefficient which for spherical particles is 2.5. Values of k_E vary according to the particle shape and orientation [4].

A large number of empirical modifications to this expression have been proposed which model the viscosity of a liquid containing moderate concentrations of spherical particles [5] These include Mooney [6], Maron–Pierce [7] and Krieger–Dougherty [8] expressions which take into account the maximum packing fraction of the particles, and where interaction effects are absent, and can be represented by the general form:

$$\eta_r = \frac{f(\phi)}{\left(1 - \dfrac{\phi}{\phi_m}\right)^{P'}} \tag{2}$$

where η_r is the reduced viscosity η/η_1, ϕ_m is the maximum packing fraction of the filler, and $f(\phi)$ and p' vary according to the different models, for example, $f(\phi)=1$ and p'=2 for the Maron–Pierce expression. ϕ_m is a function of the shape and size distribution of the particles [4].

For spheres which aggregate, immobilization of liquid between the particles increases the apparent volume of the aggregate raising the value of the Einstein coefficient k_E, depending on the number of particles present and their mode of packing [9].

Newtonian fluids containing a high concentration of rigid particles can show non-Newtonian flow behaviour with increasing shear rate, due to a break up of agglomerates in the shear field [4]. For many pseudoplastic fluid suspensions the change in apparent viscosity (η_a) with shear rate $\dot{\gamma}$ can be represented by Eq. 3 [10,11]:

$$\eta_a = \eta_\infty + \frac{\eta_0 - \eta_\infty}{1 - \Omega \dot{\gamma}^m} \tag{3}$$

where η_0 and η_∞ are apparent viscosities at very low and high rates of shear. Ω and m are empirical constants which depend on the size of the agglomerates and resistance to rupture in the shear field.

Variants to this expression and alternative models have also been proposed to describe shear thinning behaviour of concentrated suspensions [12,13].

Concentrated particle suspensions may also show a yield point which must be exceeded before flow will occur. This may result from interaction between irregularly shaped particles, or the presence of water bridges at the interface between particles which effectively bind them together. Physical and chemical attractive forces between suspended particles can also promote flocculation and development of particle network structures, which can be broken down by an applied shear stress [2].

The Casson equation is frequently applied to describe shear yielding in concentrated Newtonian suspensions [14]:

$$\tau^{1/2} = \tau_y^{1/2} + K\dot{\gamma}^{1/2} \tag{4}$$

where τ_y is the yield stress, $\dot{\gamma}$ is the shear rate and K an empirical constant.

2.3 Non-Newtonian Polymer Suspensions Containing Rigid Fillers

The rheology of unfilled polymers has received extensive evaluation and is documented in a large number of texts on the subject [15,16]. Such materials are strongly non-Newtonian under conditions experienced during melt processing. When particulate fillers are present a number of additional factors must be taken into account, including effects dependent on the additive concentration, particle shape and surface interaction with the melt [17]. Results from a number of experimental studies will be described to exemplify this behaviour at high and, in particular, at low shear rates, where shear yielding phenomena become apparent.

2.3.1 *Effect of Filler Loading*

Figure 1 shows how the viscosity of low density polyethylene-containing titanium dioxide changes as a function of apparent shear rate [18]. Similar results are shown in Fig. 2 for polystyrene filled with carbon black [19].

With each system the viscosity is shown to increase markedly with filler concentration. This relationship is particularly evident at very low shear rates and becomes even more pronounced when viscosity is plotted as a function of shear

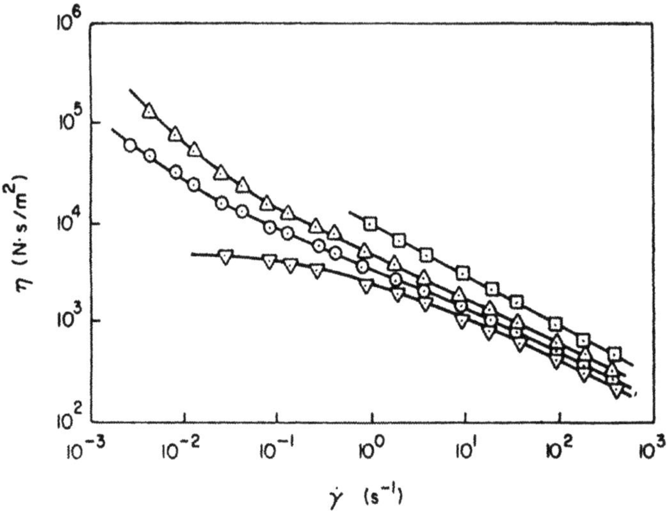

Fig. 1. The relationship between viscosity and shear rate for low density polyethylene filled with titanium dioxide (at 180 °C). Filler loading (vol%): (▽) 0 ; (⊙) 13; (△) 22; (⊡) 36 [18]

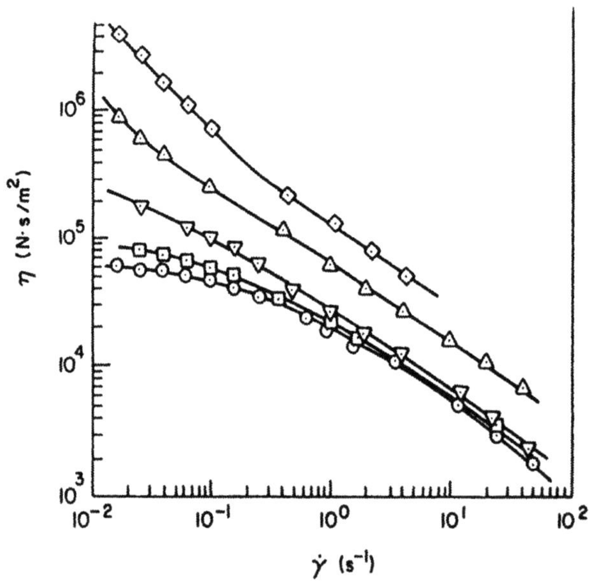

Fig. 2. The effect of shear rate on viscosity of polystyrene filled with carbon black (at (170 °C). Filler loading (vol%): (⊙) 0; (⊡) 5; (▽) 10; (△) 20; (◇) 25 [19]

stress (Fig. 3) [19]. Calcium carbonate, with a mean particle size of 2.9 μm, was found to increase the shear thinning behaviour of polypropylene as the filler level was raised up to 50% by volume [20], and is manifested by a progressive drop in power law index from 0.39 (for unfilled PP) to 0.28.

Combining very high levels of inorganic powders with organic binders provides an increasingly important route for near net shaping of engineering ceramic artifacts using polymer processing technologies, such as injection moulding. The organic phase, which may be a thermoset, or, more commonly, a thermoplastic, confers fluidity to the composition to enable shaping, then on cooling or solidification stabilises the component to resist distortion or damage during ejection and subsequent handling. In addition, the polymeric binder must be sacrificial, enabling pyrolysis and removal from the ceramic body in a non-catastrophic manner. Ceramic injection moulding compositions normally contain a very high filler volume fraction (in excess of 50 vol%), which is much higher than in conventional thermoplastics formulations, thereby imposing stringent requirements on their rheological behaviour. Hence, in addition to the ceramic powder and polymeric binder, usable moulding compositions generally contain a number of other minor constituents, including plasticisers, which function as flow modifiers, oils, which modify the pyrolysis process, and processing aids,

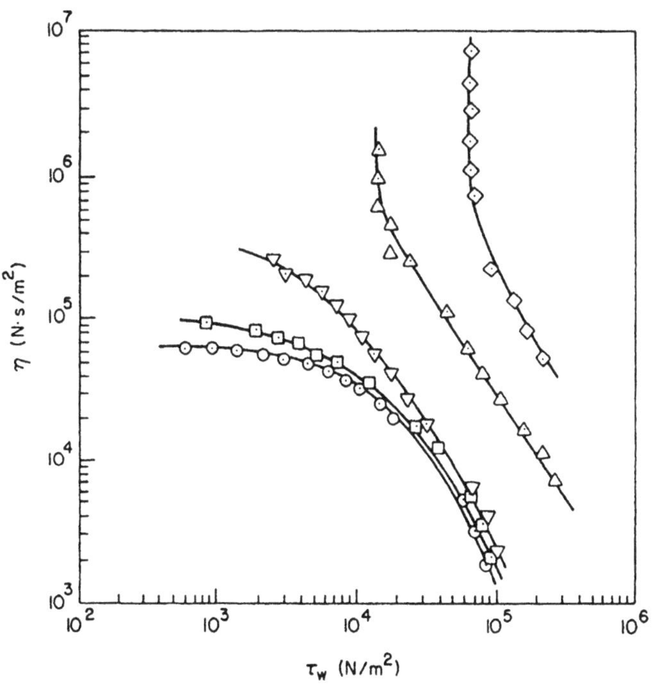

Fig. 3. The influence of shear stress on the viscosity of polystyrene filled with carbon black (at 170 °C). Filler loading (vol%): (⊙) 0; (⊡) 5; (▽) 10; (△) 20; (◇) 25 [19]

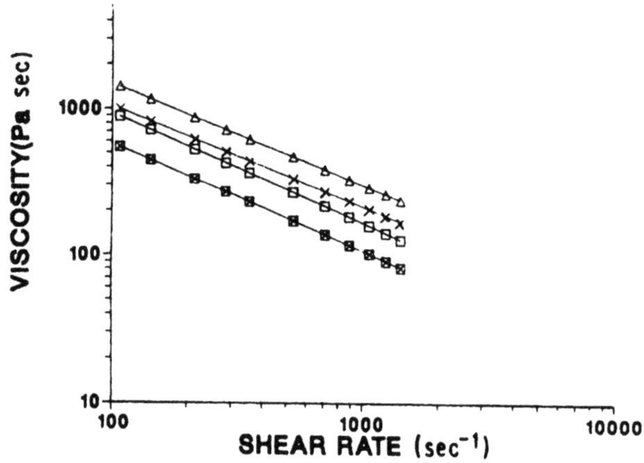

Fig. 4. The relationship between viscosity and shear rate for silicon containing ceramic injection moulding formulations given in Table 1 (at 225 °C). (△) F1; (×) F2; (□) F3; (⊠) F4 [22]

which promote particle wet-out by the binders [21]. Rheological characterization methods have been applied to the study of these highly filled polymer compositions in order to assess the interrelationship of viscosity with compositional parameters, and the dependency of viscosity on shear rate and temperature [22–24]. Figure 4 shows the effect of viscosity on shear rate for the silicon powder formulations shown in Table 1, suitable for subsequent conversion to silicon nitride after moulding and binder removal. It is clear that replacing the major binder phase by increasing amounts of secondary binder has a substantial influence on shear viscosity. Substitution of microcrystalline wax for polypropylene at a 1:7 weight ratio reduced viscosity by approximately 29% at a shear rate of 108 s^{-1}, although problems with the mechanical stability of this composition were encountered [22].

Shear thickening, or dilatancy behaviour, is commonly associated with coarse powder suspensions, such as those used in powder metallurgy [25] and with low molecular weight (wax-based) binders. This can be seen in Fig. 5 for a coarse zir-

Table 1. Ceramic Injection Moulding xFormulations [22]

Formulation	Silicon	Polypropylene (major binder)	Microcrystalline wax (minor binder)	Stearic acid
F1	82.46	15.59	-	1.95
F2	82.44	13.66	1.95	1.95
F3	82.44	11.71	3.90	1.95
F4	82.44	7.81	7.80	1.95

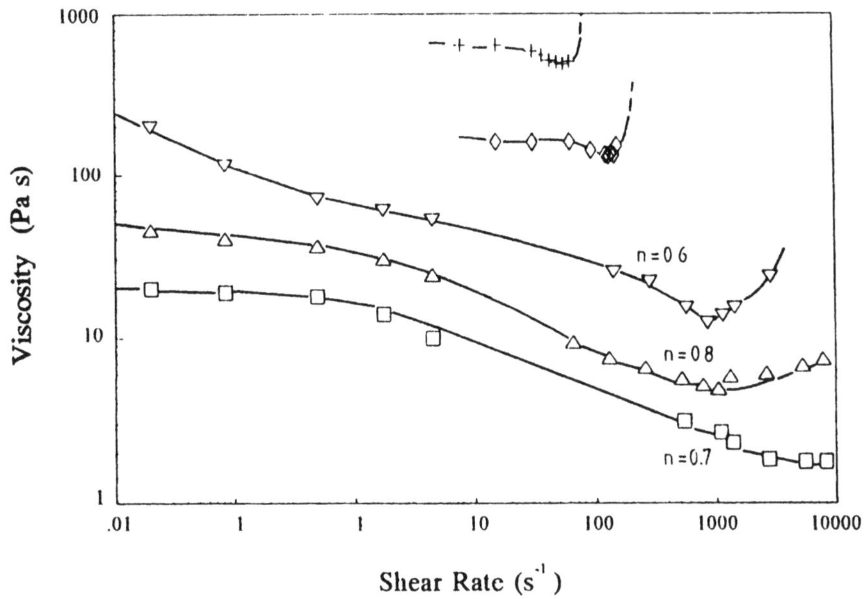

Fig. 5. The relationship between viscosity and shear rate for zirconia/wax binder injection moulding formulations (at 100 °C). n is the flow behaviour index. Ceramic filler volume fraction (%): (□) 50; (△) 55; (▽) 60; (◇) 65; (+) 70 [26]

conia powder contained in a wax binder [26]. It is apparent that the onset of dilatancy, manifested by a sharp rise in viscosity, occurs at a critical shear rate which depends on the particle volume fraction, particle size and shape and the presence of particle stabiliser. It has been proposed that shear thickening arises from a transition from a two-dimensional layered arrangement of particles to a three-dimensional network structure [27].

Using silicon nitride powder in a polypropylene/microcrystalline wax/stearic acid binder formulation, the effect of filler volume fraction (V) (over the range 50 to 70%) on relative viscosity (η_r) was predicted from Eq. 5:

$$\eta_r = \left(1 + \frac{0.75V / V_{max}}{1 - V / V_{max}}\right)^2 \tag{5}$$

where V_{max} is the volume loading at which the viscosity becomes infinite as particles make contact, or are inhibited from rotating in shear flow [28]. V_{max} is dependent on powder particle size distribution, shape and specific surface area.

2.3.2
Effect of Filler Size and Shape

The presence of fillers in viscous polymer melts not only increases their viscosity but also influences their shear rate dependency, especially with non-spherical particles (fibrous or flake-like) which become oriented in the flow field. As Fig. 6 shows, particle orientation increases the non-Newtonian behaviour which commences at a lower rate of shear than for unfilled melt.

A dimensionless relationship has been proposed to describe the bulk viscosity η $(\phi, \dot{\gamma})$ of concentrated suspensions taking into account particle size and shape effects: [29]:

$$\frac{\left[\eta(\phi,\dot{\gamma})-\eta(0,\dot{\gamma})\right]}{\eta(0,\dot{\gamma})}=F(\phi,d_p,\alpha,\dot{\gamma}) \tag{6}$$

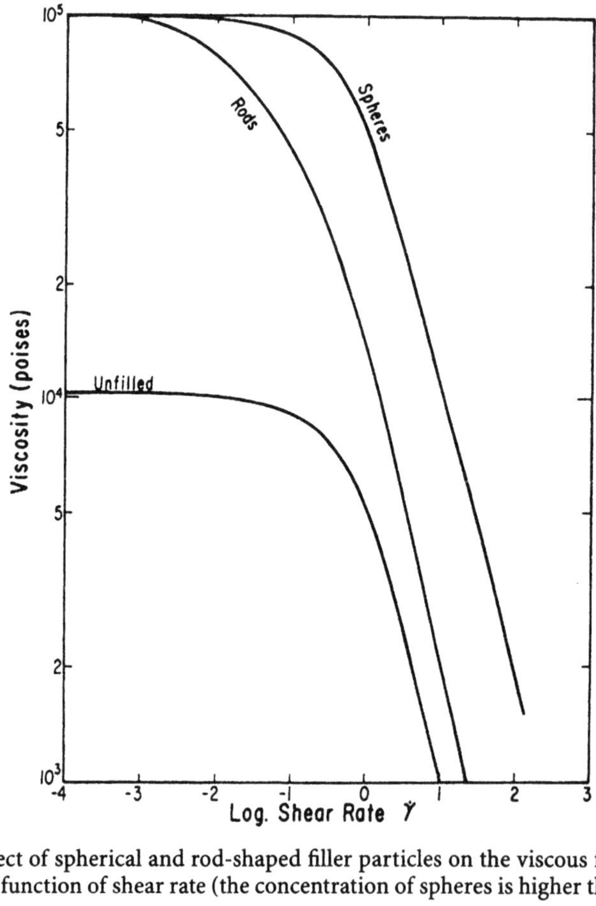

Fig. 6. The effect of spherical and rod-shaped filler particles on the viscous flow of a polymer melt as a function of shear rate (the concentration of spheres is higher than that of the rods) [4]

where $\eta(\phi, \dot{\gamma})$ is the viscosity of the suspending medium, ϕ is the volume fraction of particles in the suspension, d_p is the average particle diameter, α is a constant shape factor for the filler used and $\dot{\gamma}$ is the shear rate. This has been shown to fit experimental data for polyisobutylenes filled with carbon black [29], where viscosity was found to increase with decreasing particle size (or increasing surface area) and with filler concentration.

There has been much interest in flow and flow orientation effects with polymer melts containing anisometric particles which may be plate-like or fibrous. Flow-induced orientation of short reinforcing fibres is an area of considerable commercial importance, which is beyond the scope of the review [30].

Plate-like particles of interest in this context include mica, aluminum flake, hammered glass, magnesium hydroxide and talc. Physical properties of composites containing these additives depend strongly on the flow-induced morphology and on the distribution of residual stresses [31].

Figure 7 shows the effect of filler particle shape on the viscosity of filled polypropylene melts, containing glass beads and talc particles, of similar density, loading and particle size distribution. The greater viscosity of the talc-filled composition was attributed to increased contact and surface interaction between these irregularly shaped particles.

A number of reports discuss melt flow characteristics of thermoplastics containing mica, talc and magnesium hydroxide plate-like fillers [31–36]. Much of this work considers dynamic and capillary flow behaviour and, in particular, yield phenomena at very low shear rates. Figure 8, for example, shows, for talc-filled polystyrene, how at low shear rates the viscosity becomes unboundedly high with increasing filler level. However, at high rates of shear viscosity values converge. Using dynamic viscoelastic measurement techniques at low angular

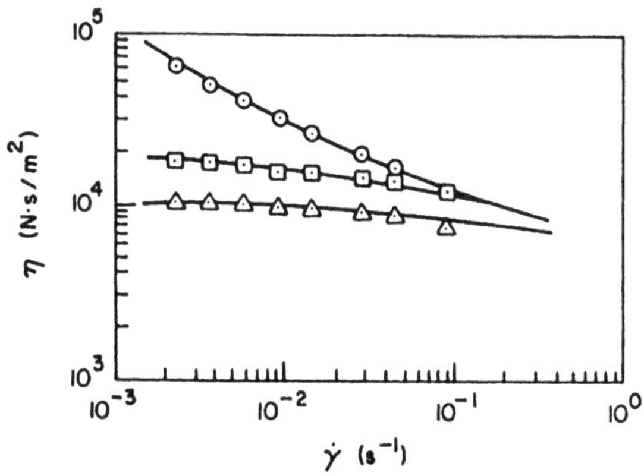

Fig. 7. The effect of filler particle shape on the viscosity of polypropylene (PP) at 200 °C: (△) neat PP; (⊡) PP containing 40% by weight glass beads; (⊙) PP containing 40% by weight talc. (Filler size distributions similar, at 44 μm or less) [17]

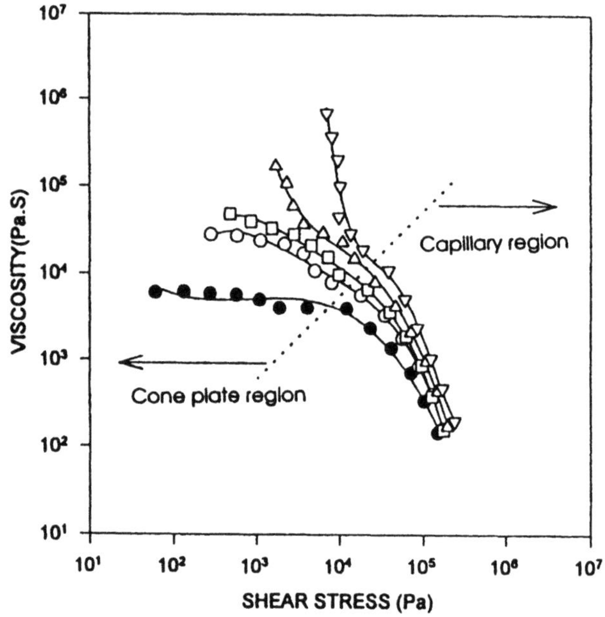

Fig. 8. The relationship between viscosity and shear stress for talc-filled polystyrene (at 200 °C). Filler levels (vol%) (●) 0; (⊙) 5; (□) 10; (△) 20; (▽) 40 [37]

frequencies, it was shown that the complex viscosity and dynamic modulus of polypropylene increased as the particle size of included magnesium hydroxide filler was decreased, i.e. its surface area increased [36].

Since the rheological behaviour of polymer melts containing anisometric filler particles during processing determines their morphology in the finished part, the interrelationship between flow and structure development has been studied in some detail. At this point, morphological investigations based on rheological experiments will be highlighted, with discussion of observed effects using specific polymer conversion technologies considered later.

It has been suggested that, in contrast to rods, particle platelets undergo a two-stage orientation process during capillary extrusion. [31,34]. Up-stream from the die they are randomly oriented, but during convergent (extensional) flow to the die they orient with the long axis in the flow direction, but randomly with respect to the capillary wall. In a simple shear field, orientation effects depend on the shear rate, filler concentration and matrix viscosity. In general, however, shearing causes disorientation of aligned particles.

In more recent work, talc-filled polystyrene compounds, with various filler volume fractions, have been processed by compression moulding and through a variety of slit, capillary, rectangular and annular dies [37]. Particle orientation has been characterised using wide angle X-ray diffraction, then expressed in the form of pole figures, and by scanning electron microscopy. It was concluded that with planar flow geometries, such as compression moulding and extrusion through rectangular dies, platelet talc particles align parallel to the mould and die surfaces. Extrusion through circular or low aspect ratio rectangular dies causes alignment of particles parallel to the die walls only at low levels of additive addition. When loading levels are high (0.2 to 0.4 volume fraction of filler), in the core of the extrudates particles exhibited a disordered radial orientation, whereas in the skin region near the die wall, they showed orientation in the circumferential direction.

Following a similar approach, rheological investigations of thermoplastics compounds containing high loadings of talc particles were compared using sandwich, cone and plate, parallel plate, capillary and elongational rheometers [35]. This study included measurements of steady state shear viscosity, transient shear viscosity, elongational viscosity and complex viscosity, and enabled measurements in shear, elongational and oscillatory flow regimes. As before, wide angle X-ray diffraction was used to analyze talc particle orientation from samples removed from the rheometers. These platelets were found to orient with their surfaces parallel to the plate surfaces in sandwich, cone and plate and parallel plate rheometers, but parallel to the flow direction in capillary and elongational rheometers. In general this work showed that the melt viscosity increased with increasing particle loading and with decreasing particle size.

2.3.3
Effect of Filler Surface Treatment

The widespread use of surface treatments for promotion of filler dispersion or enhancement of interfacial bonding between filler and polymer matrix may also strongly influence melt rheology and processability. Frequently the presence of surface modification on the filler results in a reduction in shear viscosity relative to untreated material, which may be explained by reduced interaction between the filler and dispersion medium, although a decreased tendency towards filler network formation may also be a contributory factor. If present in the polymer phase, these chemicals may also exert a lubricating or plasticizing effect causing a reduction in viscosity. There are instances, however, where surface treatment of filler can result in increased melt viscosity due to enhanced interaction between filler and polymer. This can be considered in terms of a stable adsorption layer formed around the filler increasing its effective volume [1]. The following discussion considers experimental observations which illustrate these phenomena in filled polymers modified with chemical treatments.

Polypropylene compositions containing magnesium hydroxide, with and without magnesium stearate surface treatment, were characterised at low and high shear rates using dynamic and capillary measurement techniques [36]. A significant reduction in viscosity was observed when surface treatment was present, particularly at low shear rates. In addition, with this system, the yield stress for the onset of flow was markedly reduced (Compare magnesium hydroxide variants A and E* in Fig. 9).

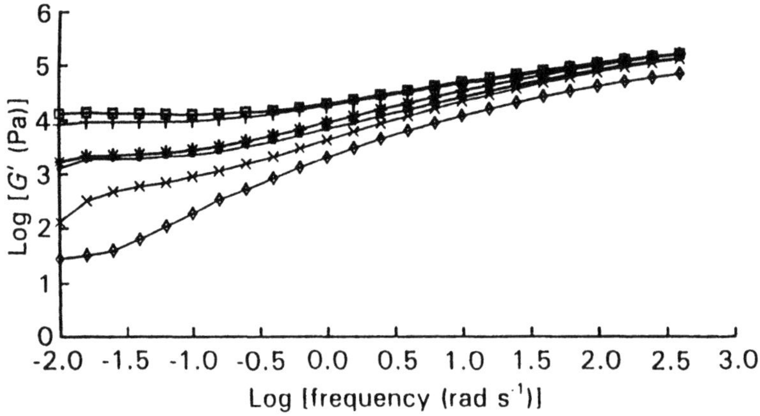

Fig. 9. The effect of magnesium hydroxide filler type on the dynamic storage modulus G' of polypropylene (PP) at 200 °C (strain amplitude 10%, filler level 60% by weight). Magnesium hydroxide fillers differed in origin particle size and treatment. Mean particle size (μm): type A (■), 7.7; type B (+), 0.9; type C (✱), 4.0; type D (□), 0.53; type E, stearate-coated version of type A, (X), 3.7; unfilled PP (◇) [36]

Small amplitude oscillatory shear experiments were undertaken in the linear viscoelastic region for mica-filled polypropylene, with and without a vinylsilane coupling agent [38]. From these dynamic measurements it was shown that an increase in mica concentration raised the storage modulus of both treated and untreated systems, although differences in their mechanisms of damping were reported. Using the concept of relative complex viscosity [39], the thickness of immobilised polymeric material on the filler surfaces was compared, since composite viscosity increased with thickness of immobilised polymer on the filler surface. It was apparent that at low mica addition levels the relative complex viscosity was higher when the filler was surface treated with silane coupling agent, indicating a greater affinity with the matrix in this system and an increase in the level of polymer adsorbed onto the filler surface.

Han [17] has shown that the effect of silane coupling agents on the viscosity of filled thermoplastics is not consistent. Melt viscosity may be decreased or increased depending on the chemical structure of the treatment and the nature of the polymer/filler combination under consideration. These observations probably reflect the effectiveness of the coupling agent in promoting bonding between filler and polymer, and hence the extent of polymer immobilization.

Shear viscosity of polystyrene containing 30% by volume of calcium carbonate differing in particle size was compared, with and without stearic acid coating [40]. The magnitude of shear viscosity and yield stress increased with decreasing particle size, but surface coating the particles significantly reduced viscosity and lowered the apparent yield values. The effects of stearic acid treatment were most pronounced with the smallest diameter particles. It was suggested that fatty acid treatment reduces the polarity of the filler surface limiting interparticle attraction.

Various investigations have considered the effects of titanate treatments on melt rheology of filled thermoplastics [17,41]. Figure 10, for example, shows that with polypropylene filled with 50% by weight of calcium carbonate, the inclusion of isopropyl triisostearoyl titanate dispersion aid decreases melt viscosity but increases first normal stress difference. This suggests that the shear flow of the polymer is promoted by the presence of titanate treatment, and is consistent with the view that these additives provide ineffective coupling between filler particles and polymer matrix [42].

The addition of carbon black to thermoplastics causes a large increase in viscosity, especially at low shear rates, where at sufficiently high additive loadings the viscosity may become unbounded due to strong interaction between the carbon black particles (Fig. 3). Surface treatments can be applied to reduce this effect and promote filler dispersion. In this regard, the mechanical and rheological properties of high-density polyethylene containing carbon black with and with-

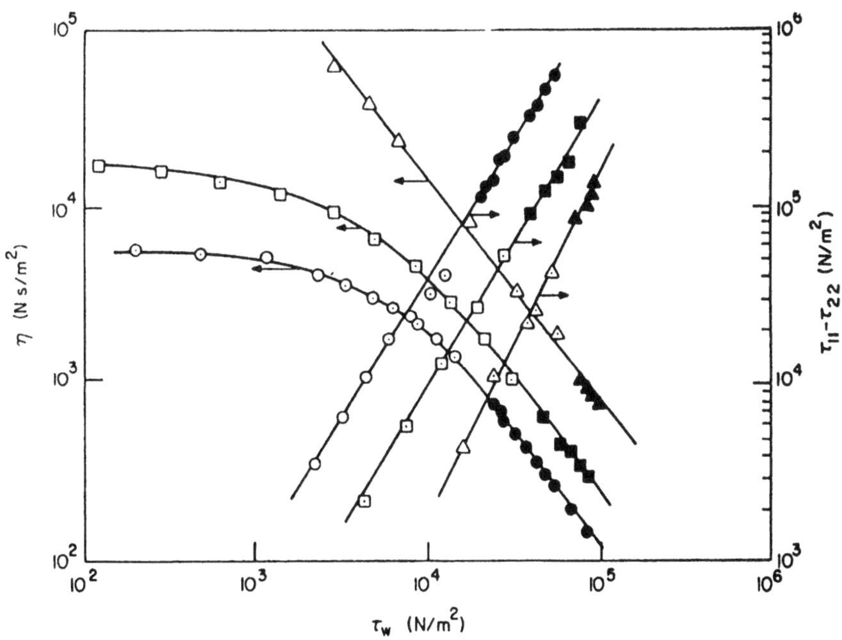

Fig. 10. Viscosity and first normal stress difference vs. shear stress for polypropylene (at 200 °C) filled with calcium carbonate (50 wt%) with and without a titanate coupling agent (TTS*): (⊙,●) pure polypropylene (PP): (△,▲) PP/$CaCO_3$=50:50 (by wt.); (□,■) PP/$CaCO_3$=50:50 with TTS (1 wt%). The open symbols were obtained from a cone and plate instrument and the closed symbols from a slit/capillary rheometer.
*=isopropyl triisostearoyl titanate [17]

out silane and titanate surface treatments were investigated [43]. It was found that the magnitude of yield stress values and viscosities were greatly reduced by the presence of treatment; however, this did not have a significant effect on mechanical properties.

Using 60 vol% of ultrafine zirconia powder (mean particle size ≤100 nm) and a wax-based organic binder formulation, it has been shown that compositions can be injection moulded through the inclusion of suitable dispersants (stearic acid or proprietary polyester material) [26]. Rheological characterization of these materials using cone and plate and capillary techniques demonstrated that the most effective dispersant used conferred a combination of high maximum packing fraction, low yield stress, the presence of a near-Newtonian plateau at low shear rate, and a high flow behaviour index in the pseudoplastic region. Coarse powder suspensions employing this dispersant showed a dilatent transition at a shear rate which depended on the filler volume fraction.

2.4 Yield Stress Phenomena

The existence of a threshold yield stress which must be exceeded for flow to occur has been indicated in earlier discussion to be a common characteristic of highly filled polymer melts, associated with interaction between the filler particles.

An overview of the origins of yield stress and parameters which can lead to variations in behaviour with highly filled polymer dispersions is given by Malkin [1]. Much of the following literature, describing experimental work undertaken, demonstrates that yield phenomena can be correlated with the extent of interaction between the filler particles and the formation of a network structure. However, the actual behaviour observed during experimentation may also depend on the deformation history of the material, or the time and temperature of imposed deformation, especially if the material exhibits thixotropic properties.

The durability of the particle network structure under the action of a stress may also be time-dependent. In addition, even at stresses below the apparent yield stress, flow may also take place, although the viscosity is several orders of magnitude higher than the viscosity of the disperse medium. This so-called 'creeping flow' is depicted in Fig. 11 where η_c is the 'creep' viscosity. In practice this phenomenon is insignificant in the treatment of filled polymer melts, but may be relevant, for example, in consideration of cold flow of filled elastomers.

Yield stress values can depend strongly on filler concentration, the size and shape of the particles and the nature of the polymer medium. However, in filled polymer melts yield stress is generally considered to be independent of temperature and polymer molecular mass [1]. The method of determining yield stress from flow curves, for example from dynamic characterization undertaken at low frequency, or extrapolation of shear viscosity measurements to zero shear rate, may lead to differences in the magnitude of yield stress determined [35].

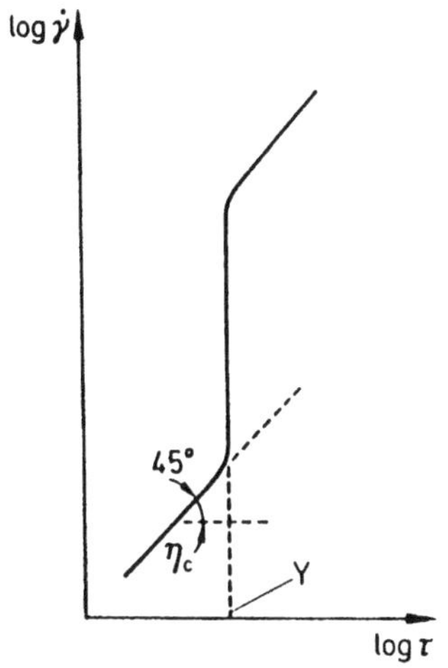

Fig. 11. The relationship between shear stress (τ) and shear rate ($\dot{\gamma}$) for a polymer disperse system showing creeping flow, with very high viscosity, η_c, at stresses smaller than the threshold yield stress [1]

Figure 8 demonstrates the phenomenon of yielding for increasing loadings of talc in polystyrene. There have been many investigations which report shear yield values for polymers through extrapolation of capillary and dynamic rheological measurements to 'zero' shear rate or stress. Using the Casson expression [14] yield values have been obtained for polystyrene and polypropylene melts filled with calcium carbonate [20,40]. The Herschel–Buckley model [44,45]:

$$\tau_w = Y + K\,\dot{\gamma}^n$$

can also be applied to the study of filled polymers which exhibit yield stresses at low shear rates, but which follow a power law relationship at high shear rates [17,35].

Several studies have considered the influence of filler type, size, concentration and geometry on shear yielding in highly loaded polymer melts. For example, the dynamic viscosity of polyethylene containing glass spheres, barium sulfate and calcium carbonate of various particle sizes was reported by Kambe and Takano [46]. Viscosity at very low frequencies was found to be sensitive to the network structure formed by the particles, and increased with filler concentration and decreasing particle size. However, the effects observed were dependent on the nature of the filler and its interaction with the polymer melt.

More recently, Lin and Masuda [47] measured the viscoelastic properties of polypropylene melts filled with small (0.15 μm) and larger (4.0 μm) calcium carbonate particles. The dynamic modulus and viscosity were found to rise with filler loading especially at low frequencies. With highly filled compositions (at

30% by weight), in the low frequency region, a second plateau appeared on the frequency-dependent curve of dynamic storage modulus G^1 (Fig. 12). This effect was most prevalent with the smaller (0.15 μm) particles and was interpreted by the formation of an agglomerated structure, which readily dispersed on application of a steady shear flow.

It has also been suggested that steady state low shear dynamic measurements in the melt could be a convenient method for the study of particle dispersion in relation to filler properties, which might also correlate with mechanical properties of the composite [48,49].

The rheological properties of gum and carbon black compounds of an ethylene-propylene terpolymer elastomer have been investigated at very low shear stresses and shear rates, using a sandwich rheometer [50]. Emphasis was given to measurements of creep and strain recovery at low stresses, at carbon black filler contents ranging between 20 and 50% by volume. The EPDM-carbon black compounds did not exhibit a zero shear rate viscosity, which tended towards infinity at zero shear stress or at a finite shear stress (Fig. 13). This was explained

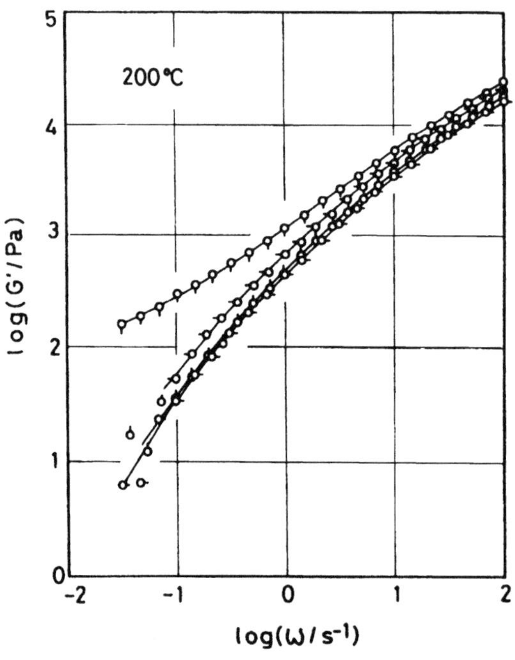

Fig. 12. The frequency dependence of dynamic storage modulus G' at 200 °C for calcium carbonate filled polypropylenes (mean particle size 0.15 μm). Filler loading wt%, (○-) 0; (ổ) 10; (-○) 20; (♀) 30 [47]

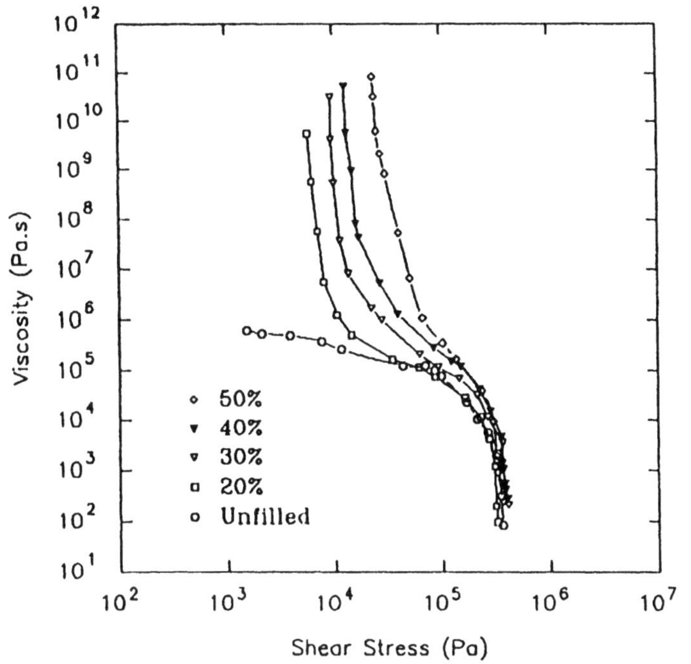

Fig. 13. Viscosity/shear stress relationship for EPDM compounds at 100 °C at various carbon black filler levels [50]

in terms of carbon black network formation. Creep and creep recovery experiments demonstrated both yield and memory behaviour with these materials.

The flow behaviour of polymeric electrophotographic toner systems containing carbon black varying in surface area and concentration were determined using a cone and plate rheometer [51]. As the concentration of carbon black was increased, the viscosity at low shear rates become unbounded below a critical shear stress. The magnitude of this yield stress depended primarily on the concentration and surface area of the carbon black filler and was independent of the polymer (polystyrene and polybutyl methacrylate) and temperature. It was postulated that at low shear rates the carbon black formed an independent network within the polymer which prevented flow.

Structure development in polymers highly loaded with carbon black has a profound effect on their electrical conductivity [52]. In this context, the interrelationship between melt flow behaviour and electrical conductivity of carbon black-polyvinyl chloride systems has been considered in terms of the extent of dispersion of carbon black aggregates, determined using scanning electron microscopy [53]. It was found that melt viscosity decreased with mixing time on a two-roll mill. During milling the carbon black aggregates changed from a cylindrical to a spherical geometry which resulted in increased dispersion in the polymer and a decrease in structure development. Electrical conductivity was found to pass through a maximum with milling time then decrease as the carbon black aggregates were broken up into fine spherical particles, thereby disrupting the electrically conductive pathway. An optimum aspect ratio of cylindrical carbon black aggregates was realised at the point of maximum conductivity (Fig. 14).

Shear yield behaviour of polymer melts containing plate-like filler particles is also prevalent and is clearly shown in Fig. 8 for talc-filled polystyrene. In this system an estimate was made of shear yield values, which were found to increase with increasing particle loading and decreasing particle size. These results are compared with reported yield values for other particulate-filled polymers in Table 2. It is evident that shear yield values also depend on the particle type and thermoplastic matrix used.

It has been suggested that the three-dimensional network structures discussed above, which are believed to occur from particle interactions at high filler loadings, may, in the case of plate-like particles, lead to anisotropic shear yield values [35]. Although this effect has not been substantiated experimentally, further theoretical interpretation of shear yield phenomena in talc- and mica-filled thermoplastics has been attempted [31,35].

The viscoelastic properties of polypropylene melts containing magnesium hydroxide fire retardant fillers have been studied using parallel plate dynamic rheology [36]. In this work the filler variants differed in particle size, surface area and morphology, ranging from approximately spherical particles formed

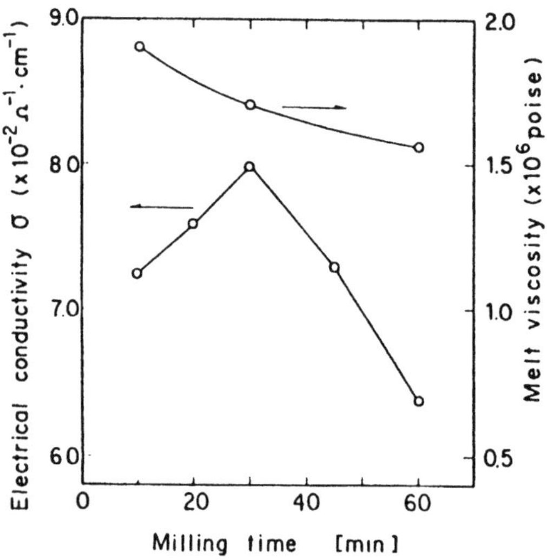

Fig. 14. Interrelationship between melt viscosity, electrical conductivity, and two-roll milling time for PVC containing 15% by weight of carbon black [53]

Table 2. Yield Values of Particle-Filled Thermoplastics [35]

Particle	Particle size (μm)	Thermo-plastic matrix	Loading (vol.%)	Temperature (°C)	Yield value (kPa)
Titanium dioxide	0.18	LDPE	25	180	2.5
	0.18	HDPE	25	180	0.5
	0.18	PS	25	180	1
	0.18	PS	30	180	22
Calcium carbonate	2	LDPE	30	200	0.12
	0.5	PS	30	180	12
	3	PS	30	180	1.5
	0.3	PS	30	180	10
	0.07	PS	30	180	40
Carbon black	0.025	PS	20	170	15
	0.25	PS	25	170	68
	0.045	PS	20	180	25
	0.45	PS	30	180	90
	0.047	LDPE	20	150	15
	0.032	LDPE	20	150	20
	0.029	LDPE	20	150	30
Talc	-	PP	18	200	0.12

from aggregated crystallites to hexagonal platelets. At a 60% by weight filler level, clear differences in response were observed between the filler types, particularly at very low shear rates. Complex viscosity and storage modulus data demonstrated the presence of a critical shear yield stress for flow to occur which increased in magnitude with decreasing filler particle size (Fig. 9).

Using model concentrated suspensions of polyvinyl chloride and titanium dioxide particles in a Newtonian polybutene fluid, small amplitude oscillatory shear and creep experiments were described [2]. It was shown that the gel-like behaviour at very small strain, and strain hardening at a critical strain, are caused by particle interactions and the state of particle dispersion.

The dynamic response of polydimethylsiloxane (PDMS) reinforced with fused silica with and without surface treatment has been discussed in terms of interactions between the filler and polymer [54]. Since bound rubber measurements showed that PDMS chains were strongly attached to the silica surface, agglomeration due to direct contact between silica aggregates was considered an unlikely explanation for the marked increase in storage modulus seen with increasing filler content at low strains. Instead three types of filler-polymer-filler association were proposed which would cause agglomeration, as depicted in Fig. 15.

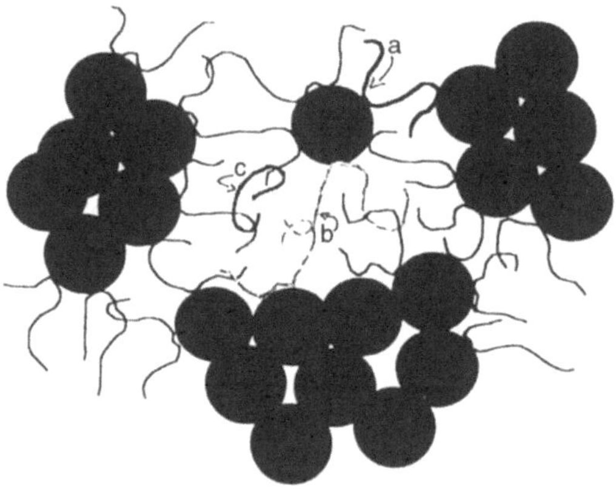

Fig. 15. Schematic diagram of different types of silica filler/polymer interaction that lead to agglomeration of aggregates: (**a**) direct bridging, (**b**) bridging through entanglement of adsorbed chains, (**c**) bridging through entanglement of non-adsorbed chains. Each group of particles in the diagram represents an aggregate (primary filler structure) [54]

2.5
Extensional Flow of Filled Polymers

It will be evident from the preceding discussion that much consideration has been given to the flow properties of filled polymers under a shear flow regime, although there is far less published information under uniaxial and biaxial extensional flow conditions. Such deformation is important in many polymer processing operations, including fibre spinning, film extrusion and foaming methods.

Measurements of the elongational flow behaviour of polymer melts containing calcium carbonate, mica, talc, carbon black, titanium dioxide and glass fibres have been reported [55–58]. Characteristic plots of elongational viscosity η_E against elongation rate η_E are shown in Fig. 16 for polystyrene containing increasing volume fractions of carbon black filler.

For the unfilled polystyrene melt at low elongational rates a constant value of η_E is achieved given by three times the zero shear viscosity η_0 according to Trouton's law [59]. However, at higher elongation rates η_E shows an increase. When filler is present η_E increases with the amount of filler, but decreases rapidly with elongation rate. Similar trends have been observed for polypropylene filled with calcium carbonate and titanium dioxide [55,58].

Extensional flow of mica-filled high-density polyethylene has been investigated in an elongational rheometer using both constant strain and stress modes [60]. Mica contents ranged from 0 to 60% by weight. Steady state elongational viscosities obtained were about ten to twenty times larger than the shear viscosity at correspondingly low rates of deformation.

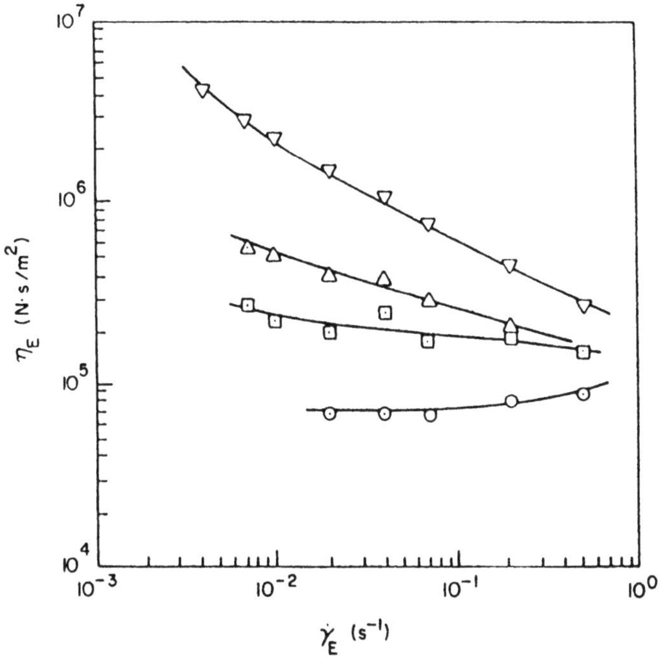

Fig. 16. The change in elongational viscosity with elongation rate for carbon black filled polystyrene at 180 °C. Filler loadings (vol%): (⊙) 0; (⊡) 10; (△) 20; (▽) 30 [17]

The elongational rheology of highly filled polyisobutylene suspensions containing 45% by volume of ceramic powder has been studied to provide a basis for understanding the flow of these materials using polymer conversion techniques, such as film blowing and blow moulding [61]. Experiments were carried out on specially designed uniaxial and biaxial rheometers. It was shown that increasing the solids content of the suspensions raised both shear and extensional viscosities. In uniaxial flow, the results followed Trouton's law for crowded suspensions containing ≤42 vol% of ceramic powder.

The phenomenon of yielding in filled polymers also exists in elongational flow but has received far less consideration than the corresponding effect seen under shear flow conditions [56,59,60]. The ratio of yield stress values obtained under conditions of uniaxial extension and shear flow have been reported between 1.4 to 1.9, which was considered to agree with von Mises criterion relating to the failure of solids at different stressed states [1], i.e. this ratio must be equal to $\sqrt{3}$ or ~1.7. It has been suggested that yielding of filled polymer melts undergoing extensional flow reduces stability during uniaxial stretching and melt-spinning processes. This is manifested by necking and low elongations to break in simple stretching experiments and periodic diameter fluctuations (or draw resonance) during melt-spinning at low draw-down ratios [62]. The transition from uniform to non-uniform stretching due to neck formation in filled polymer melts subjected to uniaxial stretching, has been shown to depend on the molecular weight of the polymer, the structural framework formed by interaction between the filler particles and changes induced during stretching [63].

3 Mixing of Fillers and Polymers

3.1 Introduction

The properties of filled polymers depend critically on the procedures used to combine them, usually by melt compounding, and the structure ultimately induced in the composition. Generally this requires that the filler is uniformly contained within the polymer and that its particle size is reduced to the minimum achievable level. The extent to which this is possible depends on many factors relating to the nature of the materials of interest, including the tendency for particles to aggregate or agglomerate and the strength of inter-particle attraction; the surface chemistry of the filler and polarity of the host matrix; and the effect on particle-matrix interaction of modifying treatments applied to the filler surface, or reactive functional groups present in the polymer phase.

Melt-mixing operations determine the extent to which particle agglomerates are ruptured and filler particles randomised within the polymer matrix. Hence an understanding of the design and operation of processing machinery in relation to levels of shear stress and shear strain developed provide a basis for optimization of the structure formation in the mixture. However, this must be considered with regard to the intended function of the additive phase or phases. For example, some fillers, such as pigments, heat and light stabilisers, are normally present in very small amounts (<5%). To maximise their performance and cost-effectiveness it is important, therefore, that they are presented in the optimum manner, necessitating high levels of dispersion and uniform distribution.

A large number of fillers are introduced into polymers at much higher addition levels, typically up to 40% by weight. Their purpose may be to alter the mechanical performance of the base material, its resistance to combustion, and electrical, optical or magnetic properties. Their mixing requirements are very much determined by the nature of the filler and its intended application. In this regard, the optical properties of pigments, including colour strength, opacity, gloss and brightness in paints, printing inks and plastics, are strongly influenced by the level of dispersion achieved. [64–68]. This in turn depends on many factors, especially the interactive forces between the pigment aggregates, the polarity of the host polymer and the intensity of the mixing regime adopted. [69,70].

In addition to its role as a pigment, carbon black may be incorporated into polymers as a reinforcement for elastomers, as a UV stabiliser in polyolefins, or as an electrically conducting additive. In each case the physiochemical properties of the filler and its ultimate state of dispersion is critical in order to achieve optimum behaviour. Again, properties such as UV absorbance are generally enhanced by increasing additive dispersion [71], although some conductive forms of carbon black are shear sensitive, requiring a balance between dispersive mixing requirements and minimal damage to the filler in order to ensure structural continuity [72,73].

A strong correlation exists between the state of filler dispersion and physical properties of filled polymer composites. With inorganic fillers, such as calcium carbonate, large particles or agglomerates present in thermoplastics can act as stress raising points, dramatically reducing tensile and impact strengths [74,75]. Similarly, processing problems have been associated with poor dispersion of reinforcing silica in natural rubber [76]. The specular gloss of polyethylene film containing titanium dioxide has also been shown to increase with filler dispersion [68] resulting from the influence of particle size on surface roughness. Brittle fibrous reinforcements and some particulate fillers, such as mica and glass microballoons, which are shear sensitive, require less intensive blending conditions to minimise additive breakage, yet ensure adequate homogeneity in the compound [77].

Organic fillers and most fire-retardant additives are thermally sensitive and may undergo decomposition depending on the time and temperature exposure during compounding [78]. The development of shear during melt blending to effect good filler dispersion may also lead to generation of shear heat arising from viscous dissipation of mechanical energy, thereby exacerbating the problem of melt temperature control.

Special requirements are necessary for the preparation of highly filled polymer composites, with additive levels at, or exceeding, 60% by weight. Pigment master batches, for example, rely on well-dispersed colourant particles which are subsequently diluted during a randomization stage during secondary melt conversion by extrusion or injection moulding. With ceramic injection moulding materials, the filler may constitute the major phase. Imperfect blending with the organic binder can, after burn-out and particle sintering operations, yield a ceramic component with potentially catastrophic internal stress raising flaws which seriously impair the integrity of the component. These may arise from improperly dispersed filler, or voids originating from polymer-rich regions due to imperfect filler uniformity in the binder [79].

In multiphase filled polymer compositions, which may contain mixed filler types, combinations of fillers and fibres, or proportions of filler and a secondary modifying polymer, such as an elastomer, the spacial distribution of the phases has a direct bearing on the properties of the composite. In the case of the last mentioned system, the rubber may encapsulate the filler, be present as discrete droplets within the thermoplastic matrix or co-exist in both structural forms [80,81].

It has been shown that the spacial location of the rubber can have a profound effect on mechanical properties [80] and may be influenced by the relative chemical affinity of rubber and plastic matrix towards the filler, the imposed shear during blending and the procedure adopted to combine the component phases, i.e. sequential or simultaneous.

The electrical conductivity of two-phase, incompatible polymer blends containing carbon black has been shown to depend on the relative affinity of the conductive particles to each of the polymer components in the blend, the concentration of carbon black in the filler-rich phase, and the structural continuity of this phase [82]. Hence, by judicious manipulation of the phase microstructure, these three-phase filled composites can exhibit double percolation behaviour.

This section will consider the fundamental aspects of melt-mixing fillers and polymers, the application of these principles to the design of polymer compounding machinery and the practical application of this and ancillary technology.

3.2 Agglomerate Formation

Finely divided fillers used as additives in polymers have a tendency to agglomerate into larger structures due to strong, inter-particle attractive forces. As mentioned earlier, it is the purpose of the dispersive mixing process to reduce the size of these agglomerates through the application of a controlled shear stress.

The structure of most particulate fillers may be classified in terms of their crystalline order and the extent to which these crystals combine together [83]. Three different species may co-exist in a powder:

- *crystals*, which may vary in size, shape and lattice structure
- *aggregates*, comprising an assembly of crystals held together by very strong forces between the crystal faces which are not generally disrupted by the dispersion process.
  Thus aggregates may constitute the so-called *primary units* in powders:
- *agglomerates*, which form where the aggregates unite, for example, at the edges or corners, to form larger and often more open structures. This assembly may behave like a single particle, but can be disrupted by considerable force.

Agglomeration of powders is an important consideration in many industries. Sometimes particles are encouraged to agglomerate to yield granules, for example, for pharmaceutical applications which may require the addition of liquids or other binders. In the ceramics, paint, plastics and rubber industries, however, reducing or eliminating agglomerate formation is of overriding importance.

The magnitude of agglomeration forces between powders depends on their surface forces and may involve electrostatic, van der Waals, or liquid-bridge forces which may occur when moisture is present [84].

In general, liquid-bridge forces are about four times larger than van der Waals forces which are at least an order of magnitude greater than electrostatic attractive forces. Adhesion forces may also depend on the contact geometry between particles, their surface roughness and their ability to undergo plastic deformation. Solid bridges may be formed by crystallizing salts or sintering, which are extremely strong and remain intact during processing with polymers.

There have been many attempts to directly measure the mechanical properties of powders which have been agglomerated by compaction [85–90].

Results from studies using calcium carbonate showed that, relative to dry powder, the effect of conditioning at 55% relative humidity (at 20 °C) increased agglomerate strength threefold, which was attributed to the development of liquid-bridge forces [86].

Measurements on the effect of various interstitial liquids (including water and polydimethylsiloxane) on the cohesive strength of various carbon black agglomerates were obtained for compacted samples with different apparent densities [91]. An order of magnitude decrease in interparticle attraction was observed relative to dry materials, which was attributed to the very high liquids content used, yielding liquid layer thicknesses larger than the particle radii. Qualitative observations using more modest levels of liquid showed that interparticle cohesive forces were significantly higher.

The effect of aggregate structure on the packing density and cohesivity of agglomerates has also been investigated using carbon black aggregates characterised by their perimeter fractal dimensions [92]. Aggregate shape was described in terms of a three-parameter model which enabled the effective size of interacting aggregates and contact points within agglomerates to be estimated. The cohesivity of agglomerates formed from aggregates were measured using a tensile test method and the results interpreted in terms of Hamaker constants, which expressed the extent of cohesive interaction.

Formation of agglomerates by powder compaction may involve rearrangement of particles to increase their packing efficiency resulting in the enhancement of interparticle adhesion forces [89]. Furthermore, particle deformation at the point of contact between particles can greatly increase both the contact surface area and interparticle attraction [84].

Compressive forces experienced by fillers during combination with polymers may also result in increased agglomerate formation. For example, it has been shown that large pigment agglomerates observed on the rotor blades of a high-speed powder pre-mixer are as hard and strong as those produced by tableting [93]. There is also evidence to suggest that fillers such as calcium carbonate may agglomerate during the early stages of extrusion compounding with thermoplastics, where filler is compacted prior to polymer melting [94].

3.3 Fundamentals of Polymer Mixing

In fundamental terms, it is helpful to distinguish between two different mixing mechanisms - extensive and intensive (or dispersive) [83,95].

When mixing very viscous polymeric systems *extensive* mixing is achieved largely by convection, which may be distributive or laminar, with the overall aim of bringing about compositional uniformity.

Distributive mixing relates to rearrangement of the components through an ordered or random process, such as in the pre-blending of polymer components in the solid state as in tumble or high-speed mixing operations. Laminar mixing, however, is achieved by subjecting the material to permanent deformation in various laminar flow patterns involving shearing, squeezing or elongational flow. This will necessitate mixing in the melt state through the imposition of large strains and is generally accompanied by an increase in interfacial area between components in the system. Effective laminar mixing is strongly influenced by the initial orientation and spacial location of interfacial elements (or solid additives) [96].

Intensive or *dispersive* mixing normally involves rupture of agglomerates formed by a solid phase. Within a polymer melt, dispersive mixing of a minor particulate phase is achieved by localised application of shear stress. Breakdown of the structure is accompanied by distribution of the separated particles throughout the polymer matrix through an extensive mixing step. Dispersive mixing of particulate-filled polymers is therefore influenced by a variety of factors relating to machine design and operation, together with material composition.

There are several approaches to the assessment of mixture quality in polymer-based compositions, including statistical analysis of the phases using the concepts of variance and correlation to describe the extent of homogeneity and spacial structure, respectively [96,97]. These characteristics can be defined in terms of the *intensity* and *scale* of mixing and may be applied to filled polymers providing that the phases can be identified. In this regard, with filled thermoplastics, combined use of microscopic and image analysis techniques can provide an effective means of characterizing dispersive and distributive mixture quality [98,99]. Within the context of laminar shear mixing *striation thickness* between phases can provide a useful measure of mixedness where the concept of striations is meaningful, for example, in pigment-filled polymers [100]. Through this approach, the extent of mixing can be related to the applied shear rate and shear strain experienced by the material during mixing [101].

Measurement of residence time distribution (RTD) through continuous flow systems can provide a measure of a material's shear and heat history, expressed in terms of a distribution function [102] and thereby yield a quantitative measure of the mixing capability during processing. Various tracer techniques, including neutron activation analysis and ashing methods, have been developed to characterise RTD in continuous polymer extrusion machinery [103]. From knowledge of the residence time distribution and shear rate of fluid elements passing through processing machines, their effectiveness as mixing devices can be determined. Expressed as the weighted average total strain (WATS), this may be related to extruder design and operating parameters [104].

Combining particulate fillers, such as carbon black, into thermoplastics involves the following sequential, but to some extent, overlapping stages [83]:

(i) **Filler Wetting.** Polymer wets the filler and squeezes into its void spaces, such that loose filler particles disappear and air introduced into the compound by entrapment in the filler agglomerates, is replaced. Polymer is believed to penetrate not only the void space of the agglomerate, i.e. between the aggregates, but also within the aggregates. It has been proposed that in a carbon black/rubber system, rubber which fills the void space within each aggregate is occluded and immobilised, thus acting as part of the filler rather than a component of the deformable matrix [105].

(ii) **Dispersion.** This relates to break up of the agglomerates and separation of the resulting fragments to a point where reagglomeration will not occur.

It is generally accepted that agglomerates will break when internal stresses, induced by viscous drag on the particles, exceed a certain threshold value. This basic concept has been developed and modelled over many years, McKelvey [106], Dizon et al. [107], and by Tadmor [95, 108].

Figure 17 depicts the suspension of a carbon black agglomerate in a viscoelastic medium. The application of a shearing strain will generate a hydrodynamic drag force acting on the agglomerate tending to separate it at its weakest link. Through a force balance, the following equation is obtained:

$$K = 6\pi \, Re \mu \dot{\gamma} / Fa \tag{7}$$

where K is the dispersibility factor, Re the agglomerate radius, μ the matrix viscosity, the strain rate and Fa the interaggregate cohesive force.

In the agglomerated stage, when Re is large, the hydrodynamic drag overwhelms the interaggregate cohesive force and dispersion proceeds efficiently. As Re gets smaller, the force balance becomes increasingly less favorable for dispersion until a certain equilibrium is reached between the two opposing forces and further dispersion is no longer possible.

This analysis was extended to pairs of interacting particles in shear and extensional flow, by considering the forces acting on a single agglomerate in the form of a rigid dumbbell comprising two unequal beads of radii, r_1, and r_2, whose centres are a distance L apart, in a homogeneous velocity field of incompressible Newtonian fluid [108]. Due to viscous drag on each of the beads, a force develops in the connector, which depends on the magnitude of the viscous drag and on the dumbbell orientation. When this force exceeds a critical value, equal to the attractive cohesive force, the beads break apart. In simple *shear* flow, the

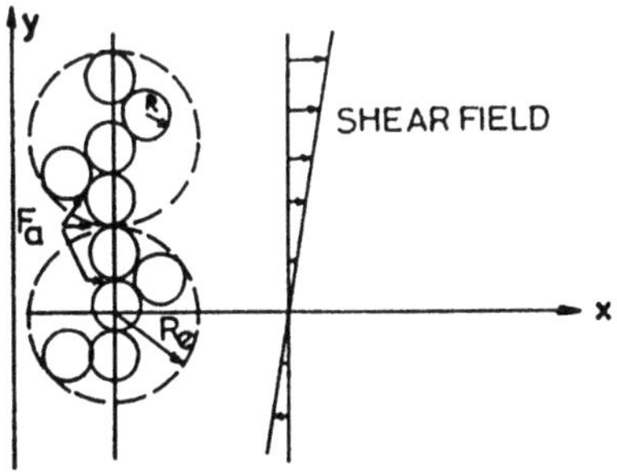

Fig. 17. Interactive forces in the dispersive mixing of carbon black. R, aggregate effective radius; R_e, agglomerate effective radius; F_a, interaggregate cohesive force [83]

maximum separating force in the connector will be obtained when the dumbbell is placed in the x-y plane at an orientation of 45° relative to the direction of shear. For the special case of two beads in contact this force (F_{max}) is given by:

$$F_{max} = 3\pi\mu\dot{\gamma}r_1r_2 \tag{8}$$

However, in steady *elongational* flow, the maximum separating force in the connector is obtained when the dumbbell is aligned in the direction of flow and, again, for the case of two beads in contact is given by:-

$$F_{max} = 6\pi\mu\dot{\epsilon}r_1r_2 \tag{9}$$

where $\dot{\epsilon}$ is the rate of elongation.

This model predicts that maximum separating forces in extensional flows are twice those in shear flows at the same deformation. Unfortunately, laminar shear is the dominant flow mechanism in most polymer compounding operations, such as twin-screw extrusion. Furthermore, it is evident that larger contacting particles are easier to separate than smaller ones and that increasing melt viscosity, e.g. by increasing polymer molecular weight, increasing filler loading or decreasing melt temperature, will also facilitate dispersion by increasing F_{max}.

The main drawback of these models is that they lack any fundamental description of the internal cohesive forces resisting rupture and fail to consider any time dependency in the dispersion process.

With reference to dispersive mixing of carbon black in a Banbury-type internal mixer, agglomerate break up has been modelled as a repetitive process, resulting from multiple passage through high shear zones in the mixer [83]. Hence, an initial large agglomerate ruptures first into two equal-sized fragments, after which each fragment is considered as a new agglomerate of smaller size which may be broken again in a following pass. The process continues until the ultimate particle size can no longer be broken by hydrodynamic forces existing in the system. A limiting degree of dispersion is achieved where agglomerates above a critical size are considered undispersed.

Hence, fragment separation becomes a dominant factor in the dispersive mixing process, and has been modelled in four flow fields, viz. simple shear flow, pure elongational flow, uniaxial extensional flow, and biaxial extensional flow [109]. The efficiency of each of these flow fields in dispersing solid agglomerates was compared on the basis of time and power requirements for a given degree of mixing, where biaxial extensional flow was found to be most efficient.

Studies on the kinetics of carbon black dispersion in various rubbers have been reported using a Brabender mixer fitted with cam-type rotors [110]. Dispersion rating, determined by visual inspection of photomicrographs, was found to depend strongly on mixing time. For an SBR emulsion, it was observed that there was an initial delay period where the carbon black agglomerates were thought to be fractured and incorporated into the rubber. Subsequently, the process of dispersion continued for a considerable time thereafter.

A specially designed dispersion tester has been constructed in order to systematically study isolated variables considered important during melt mixing using carbon black in SBR rubber [111]. The results confirm that the number of passages experienced by material through the high shear region is a dominant variable in the dispersive mixing process, in addition to the stress history of fluid elements.

The extent and mechanism of carbon black structural breakdown during polymer mixing has been an area of some controversy in the literature. Some earlier reports have provided evidence for significant aggregate breakdown during mixing into elastomers [112]. Other workers, however, have suggested that there is no permanent breakdown of aggregate structure and that this represents the finest state of subdivision into which carbon black can be dispersed [113].

More recent work has provided a clearer insight into the mechanism of carbon black dispersion during simple shear flows [114,115]. Experiments were undertaken in a transparent cone and plate device using polydimethylsiloxanes with a wide range of viscosities, as suspending media for aggregated carbon black in pellet form. Two distinct break-up mechanisms were observed, denoted as 'rupture' and 'erosion'. The rupture process is characterised by an abrupt splitting of agglomerates into a small number of large fragments (rather than by splitting into two halves as was seen in earlier work) and tends to occur at relatively high shear stresses.

The presence of weak spots and structural non-uniformities in the agglomerates were considered to be determining factors in the break-up process [116,117].

The erosion process is more gradual, however, and initiates at lower applied shear stresses than rupture, being characterised by the progressive detachment of small fragments from the outer surface of the agglomerate. For the erosion of carbon black agglomerates suspended in Newtonian fluids, it was found that the size of the eroded fragments obeys a normal distribution and that the kinetics of the process follow a first-order rate equation of the following type:

$$\frac{R_0 - R_{(t)}}{R_0} = Kt^* \tag{10}$$

where R_0 is the initial cluster radius and $R_{(t)}$ the radius after time t; t^* is a dimensionless erosion time given by the product of t and applied shear rate ($\dot{\gamma}$). K depends on flow geometry, applied shear stress and cohesive strength of the agglomerate.

This is somewhat similar to the 'onion peeling' mechanism of dispersion first proposed by Shiga and Furutu [118] which involves a scraping action of the moving matrix at the surface of the agglomerate, causing individual constituent particles to be removed from the agglomerate. However, since carbon black agglomerates have non-uniform structures, it has been postulated that small asperities projecting from the surface are weak and break off under hydrodynamic stresses [114]. Hence a steady state would be reached at which most agglomerates had lost all of their weak asperities, leaving the strong (and relatively large) cores, in addition to a large number of very small particle fragments.

Analysis of the dispersion of carbon black agglomerates in polystyrene and high-density polyethylene melts has confirmed that coarse rupturing of agglomerates occurred in the early stages of dispersion, followed by more gradual erosion of small aggregates from larger fragments. [71].

However, controlled shearing experiments using titanium dioxide in PDMS and linear low density polyethylene demonstrated that with this filler type, particle erosion was the predominant dispersion mechanism [68,119].

A model has been developed to describe the penetration of polydimethylsiloxane (PDMS) into silica agglomerates [120]. The kinetics of this process depend on agglomerate size and porosity, together with fluid viscosity. Shearing experiments demonstrated that rupture and erosion break-up mechanisms occurred, and that agglomerates which were penetrated by polymer were less readily dispersed than 'dry' clusters. This was attributed to the formation of a network between silica aggregates and penetrated PDMS, which could deform prior to rupture, thereby inhibiting dispersion.

(iii) **Distribution.** Once the filler agglomerates are broken, separation of the closely spaced agglomerate fragments and their distribution throughout the polymeric matrix should be accomplished. Accordingly, the degree of filler distribution no longer depends on stress, but only on the total level of strain applied to the matrix.

It has also been suggested however, that even with well-separated aggregates of carbon black, flocculation may occur by diffusion in a hot rubber compound, contributing to the formation of a network and an electrically conducting structure.

Coalescence is another reverse process in mixing. Whereas this is a major issue in the formation of polymer blends, it is considered of less significance with carbon black or other solid filler dispersions in polymers [83].

4 Polymer Compounding Technology

4.1 Introduction

The successful incorporation of fillers into thermoplastics relies on the efficiency of the combining procedure and the quality of the mixture produced. Blending operations are normally effected with polymer in the melt state, but low intensity pre-mixing operations involving additive(s) and solid polymer are also common, particularly when this feedstock is in powder form. Hence the design of compounding machinery must be sufficiently flexible to accommodate the processing requirements of both filler and polymer in terms of their necessary heat and shear history at a specified throughput rate. Compounding plant may be batch or continuous in design, but the usual product from this operation is a pellet of uniform composition and geometry suitable for use in subsequent injection moulding or extrusion die-forming operations.

Various forms of ancillary equipment are required, in particular to ensure consistent addition of feedstock streams to the process equipment, to undertake melt cooling and pelletisation, and, in some instances, to filter out extraneous contaminants.

A diverse range of equipment types are available commercially for compounding plastics and rubbers. In most instances their design has evolved from empirical considerations and the need to cater for increasingly demanding compounding situations. More recent attempts to model compounding machinery is providing the basis for further optimization of process design and operation, although a large gap still exists between theory and practice, especially with more complex techniques such as twin-screw extrusion.

An account will be given of established approaches which are used to incorporate fillers into thermoplastics, together with developments in this field and an appraisal of specific measures which must be considered in the practical operation of this technology to prepare filled thermoplastics composites.

4.2 Process Requirements for Compounding Filled Polymers

Modern melt-mixing machinery has all or most of the following functional capabilities depending on whether it operates using batch or continuous principles.

(i) **Transport of Feedstock and Melt Compound.** During continuous operation it is necessary to regulate the flow of materials into the process chamber consistently and in the specified proportions. Commonly, this necessitates use of accurate dosing equipment, for example, when metering filler and polymer into twin-screw extruders either at the same feed position or at separate addition ports. Conveying of unmelted polymer and filler relies on frictional differences between polymer and metal surfaces creating a so-called drag flow mechanism evident in single-screw extruders, or positive displacement of material through interaction of mechanical features, such as intermeshing screws. In some instances, both mechanisms may occur simultaneously [95]. Mixtures of solid polymer and high levels of filler may markedly influence frictional coefficients and hinder conveyance by drag flow. In such circumstances use of a grooved feed barrel section or preferably a closely intermeshing twin-screw extruder may overcome this limitation [121].
At the output end of the compounder, removal of molten compound may be as a large mass via a discharge port from an internal mixer, or through a die which defines the geometry of the compound. In the latter case, sufficient pressure must be developed to overcome the flow restriction imposed by the die.

(ii) **Melt Generation.** The energy required to convert polymer from a solid to a melt state during compounding may be derived in part from thermal conduction from the boundary of the mixer, but more significantly from dissipation of mechanical energy in the form of shear heat. Although this can arise from friction between solid particles, shear heating can be extensive during the process of polymer melting, both within molten material and especially at the solid melt interface [104]. Clearly, factors which affect the rheology of the polymer, such as temperature or the presence of filler, influence the extent of shear heat generated. In this regard, an increase in melt viscosity will occur, which will depend both on the nature and level of filler present and also on its specific heat capacity, leading to a reduction in localised temperature around the filler particles.

(iii) **Mixing Phenomena.** The combination of filler and polymer melt was considered at length earlier in terms of convective and dispersive mixing processes. In the context of compounding machinery design, mixing is effected by the judicious interplay between operating parameters (temperature/pressure generation/mixer speed) and design features, which influence randomization of the mixture and controlled passage through clearances, thereby defining the magnitude of applied shear stress and overall shear strain developed. These parameters characterise the essence of polymer compounding operations and can be varied by a variety of ingenious methods, providing the process flexibility necessary to accommodate different formulation requirements.

(iv) **Melt Devolatilization.** The quality of filled polymer compounds can be greatly enhanced through provision of melt devolatilization procedures. This is very common practice in both single- and twin-screw extrusion compounding to facilitate removal of entrapped air, moisture associated with filler or polymer phases, or reaction byproducts which can arise during chemical treatment or polymer modification processes. The normal requirement is to design a low-pressure decompression zone into the process chamber to enable volatiles to be removed at this point under vacuum. Many different variations exist, however, and through analysis of the principles governing volatile diffusive and convective mass transport, machine design and operational variables can be optimised [122].

4.3 Constructional Features of Compounding Equipment

A large number of polymer compounding variants are available for the preparation of filled polymer compositions. In commercial practice, machine development has focused on providing greater operational flexibility, improvements in product quality and consistency and an ever-increasing drive towards higher throughput. Although batch mixing equipment continues to be in widespread use, particularly in the rubber industry, in many areas of thermoplastics compounding, continuous extrusion processing technology is the preferred option. There is also a trend towards integration of polymer compounding and end-forming technologies which can offer both technical and economic opportunities. Reactive compounding technology is in widespread use for the chemical modification of polymers and can also be pertinent to blending of filled polymer formulations.

It is the intention to consider the design features of polymer compounding machinery in terms of their functional capabilities, as identified previously, specifically in relation to the processing of filled polymer compositions.

4.3.1
Pre-Mixing Procedures

Low and medium intensity blending procedures are primarily applied to pre-distribute fillers and polymers prior to more intensive melt compounding. Although their application may encompass liquid/solid mixtures, including PVC paste and polyester-based formulations, such as dough moulding compound, in the context of the present discussion their sole function is very often simply to randomise the spacial distribution of polymer and additive particles, such as pigments or inorganic fillers. Segregation of the components is a common problem, due to differences in particle size and density, which can be ameliorated using mixtures of powdered polymer and filler, or promoting additive adhesion to polymer granules through incorporation of wetting agents.

Many basic mixer variants exist, including ribbon blenders, V-blenders, conical screw mixers and planetary mixers, which differ in operational complexity, efficiency and the capability to combine feedstock with differing physical characteristics [123]. Pre-mixing techniques can be used in isolation, prior to melt compounding or integrated into this stage, for example, by blending components at the point of dosing into the feed throat of a compounding extruder.

In general, mixers of the type outlined above impart low shear and hence do not significantly influence the physical size of agglomerated filler particles. So-called *high intensity non-fluxing batch mixers* increase the level of applied shear, but without melting the polymer. Their principle application is in the preparation of PVC dry-blends where modifying additives, such as stabilisers, lubricants and plasticisers, are adsorbed onto powdered polymer. This is achieved within a mixing chamber containing an impeller which can rotate at speeds up to 4000 rpm (Fig. 18). This causes the contents of the mixer to fluidise resulting in a high degree of particle randomization and generation of a significant temperature rise due to shear heating. The operation of the mixer and conditions experienced by the PVC blend have a significant influence on subsequent processibility [124].

High-speed mixers can also be highly effective for intimate blending of inorganic fillers, or pigment particles, with powdered polymer and for application of surface treatment onto filler surfaces. Silane and titanate treatments may be conveniently drip fed from solution onto the fluidised filler by this method, enabling controlled and uniform coating [125]. With fatty acid surface treatments most effective surface coverage onto magnesium hydroxide was found by introducing surface treatment onto pre-heated filler (above the treatment melting temperature) within the mixer, rather than blending at room temperature [126].

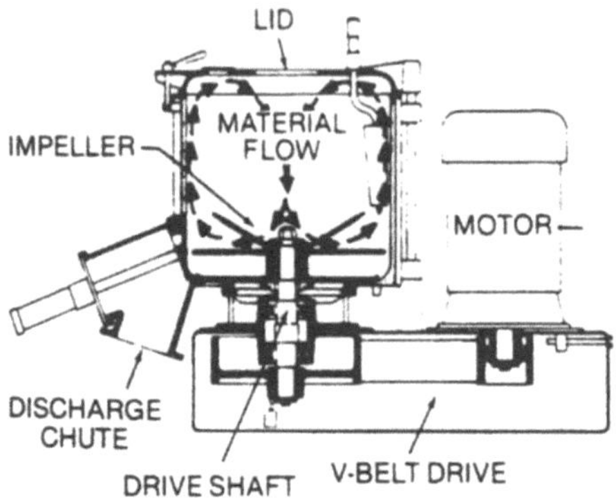

Fig. 18. Schematic diagram of non-fluxing high-speed mixer

In this context, diffuse reflectance FTIR spectroscopy has been applied to analyze the influence of mixing variables (i.e. treatment concentration, time and temperature of mixing) on both the ultimate filler coating level and possible reaction between the treatment and filler surface [127].

There is evidence to suggest that the intense action created inside the chamber of a high-speed mixer is sufficient to cause compaction and agglomeration of finely divided fillers [93], or damage to shear sensitive additives, such as glass fibres or naturally occurring fillers derived from wheat straw [128].

4.3.2
Melt-Mixing Technologies

Processes which melt mix polymer and filler are capable of generating the high shear stresses necessary to cause agglomerate break up, together with re-distribution of the primary filler particles. Since its conception in 1835, the *two-roll mill* has proved to be an effective means of mixing additives into plastics and rubber and is still in widespread use today, principally for laboratory purposes, but to some extent in large-scale industrial applications.

As shown in Fig. 19, polymer adheres to one of the two counter-rotating rolls passing through an adjustable nip gap which creates intense shear. The rolls are temperature controlled and may run at differential speeds to influence further shear intensity. A degree of lateral cross-mixing must also be imposed to ensure overall compositional uniformity of the batch. A rolling bank of polymer located above the rolls provides some additional mixing capability.

Two-roll mills have been analyzed in terms of the pressure distribution and velocity profiles created between the rolls [95], the shear imposed on fluid elements exposed to these conditions in the nip region [129] and their resulting efficiency as dispersive mixing devices [130,131]. An earlier mathematical model was proposed to describe the dispersive mixing process of carbon black in rubber on roll mills, through consideration of agglomerate size distribution and

(a)

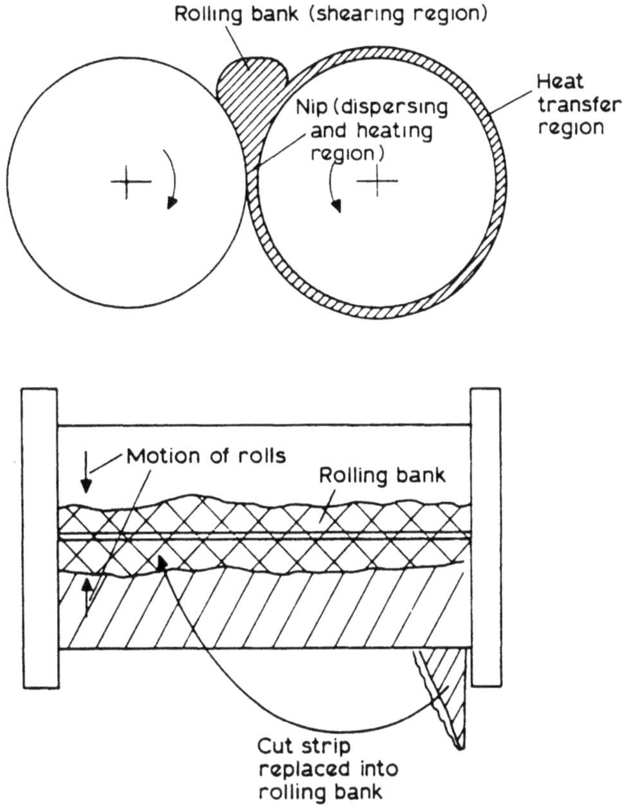

(b)

Fig. 19. Two-roll mill showing: (**a**) cross-sectional view of rolls, rolling bank and nip and (**b**) plan view of rolls and convective effect of manual intervention [129]

process parameters, including nip gap size, friction ratio between the rolls, roll speed and mixing time. [130]. The model was based on the assumption that intensive mixing is dominated by agglomerate rupture in the high shear field existing between the rolls and did not consider the influence of the re-circulating region at the entrance of the nip, which may also contribute to composition uniformity. However, more recent analysis of the fluid dynamics within the bank and nip area has been reported using a finite element method [131]. Dispersive mixing efficiency was described in terms of the shear stresses generated and elongational flow components. It was concluded that the converging region rather than the actual nip zone has more influence on mixture quality, and that flow vortices present in the bank region do not benefit dispersive mixing. Furthermore, increasing friction ratio between the rolls does not improve overall mixing performance.

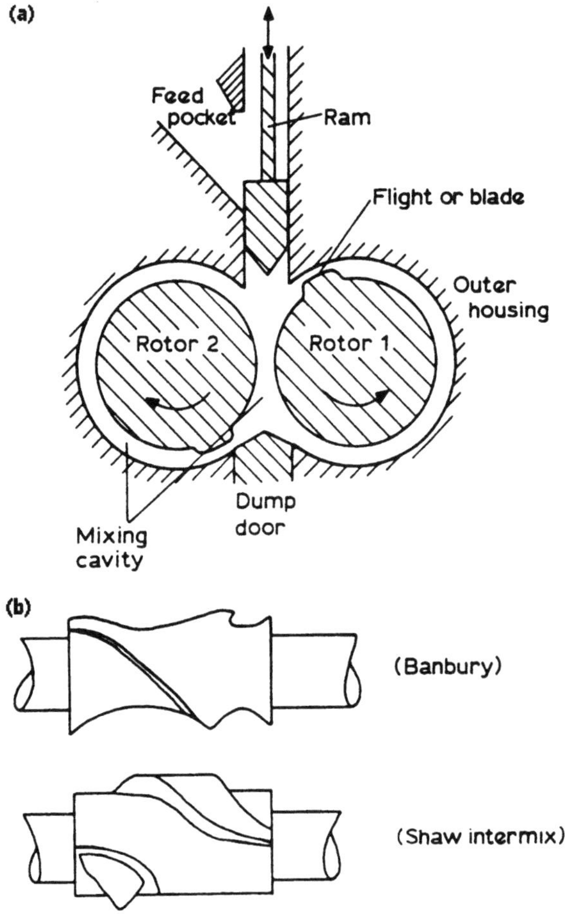

Fig. 20. Internal mixer design showing: (**a**) cross-sectional view of constructional features and (**b**) typical rotor geometries [129]

Limitations of the two-roll mill can be overcome in batch and continuous forms of *internal mixer.* A typical mixer design, shown in Fig. 20, comprises a closed temperature-controlled chamber in the shape of a figure of eight containing two counter-rotating intermeshing rotors. The polymer composition is fed to the mixing chamber through a vertical chute which also houses an air or hydraulically driven ram. When lowered onto the feedstock this creates pressure and intensifies the mixing action. On completion of the mixing cycle material is discharged through a door at the base of the mixing chamber, often onto a two-roll mill or into a melt extruder, for subsequent pelletization. The rotor design, together with mixing variables, including rotor speed, batch temperature, ram pressure fill factor and loading pattern, all determine the extent and mechanism of mixing within the process chamber. For example, during rubber compounding feeding procedures include 'dump mixing' where all the ingredients are added simultaneously, upside down mixing, where solid additives are introduced before the polymer, conventional mixing, where the rubber is loaded first followed by other ingredients together or in stages after a prescribed time period, and seeding, where a quantity of previously well-mixed compound is added to the new batch [95]. Rotor designs can differ in profile and whether or not they intermesh. Small clearances exist between the rotors and chamber wall and in some instances between the rotors themselves, where high shear stresses are generated. The rotors should also ensure that material is uniformly transferred throughout the mixing chamber (Fig. 21). A large number of studies have been undertaken which consider the mixing of polymer within internal mixers in relation to rotor designs. These include flow visualization studies [133–135] and effects from changing processing parameters [136]. Many experimental and theoretical reports, typically using finite element analysis principles, have analyzed mixing and operational performance of internal mixers differing in rotor design, for example, between intermeshing and non-intermeshing variants [137,138], or the number and geometry of rotor wings [139–144].

Much of this analysis is aimed at understanding the nature of velocity profiles, shear and elongational flow behaviour of fluid elements experiencing the unique mixing action within internal mixers. Due to their importance in the rubber in-

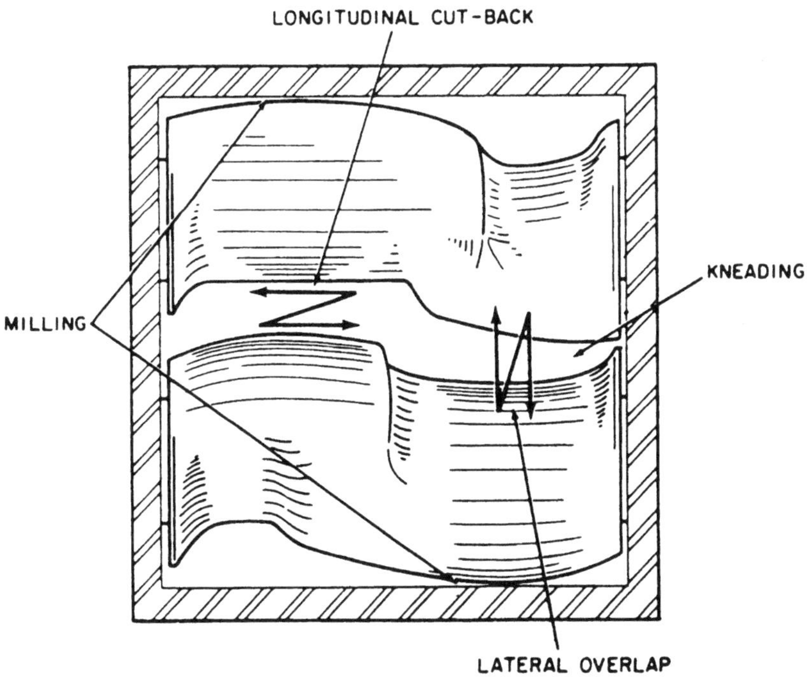

Fig. 21. Mixing mechanisms in an internal mixer [132]

dustry, it is not surprising that reports on dispersive mixing capability have largely focused on using elastomeric formulations reinforced with carbon black and silica fillers, although these accounts also provide an insight into specific criteria influencing dispersive mixing using alternative polymer/filler combinations [145].

The mixing principles of internal mixers mentioned above are also evident in continuous forms of this device. Feedstock is introduced into the mixing zone using counter-rotating twin screws and the mixture progressively metered through a discharge gate [146]. Commonly, material processed by the mixer is delivered into a single-screw extruder to allow some further homogenization, possible venting, then pressurization to die-form the compound.

Two-stage rotors are also available for continuous mixers (Fig. 22), which incorporate a second screw section to enable down-stream feeding of additives such as heat or shear sensitive fillers, or extraction of volatiles from the mixture [147]. For a fixed rotor geometry, rotor speed, flow rate, barrel temperature and orifice opening are the principal operational variables which control mixing intensity.

Preparation of thermoplastics compounds containing particulate fillers is dominated by the use of *continuous screw extruders* which are available com-

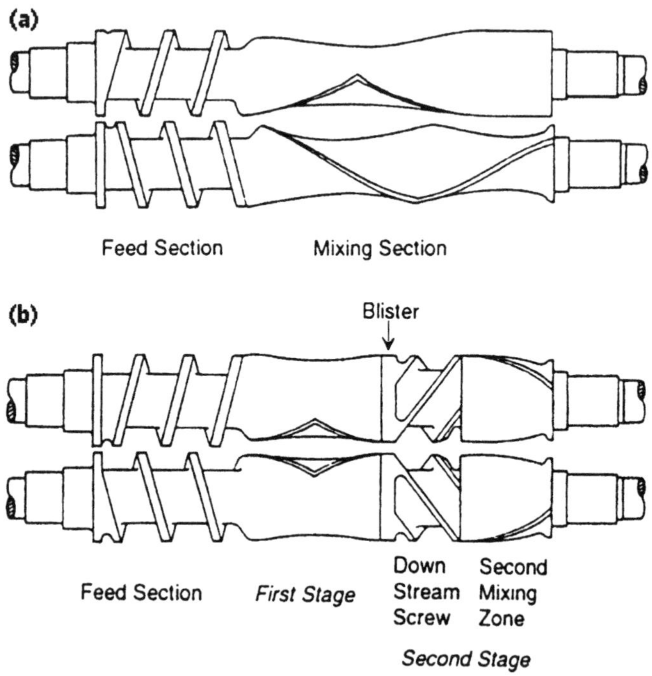

Fig. 22. Rotor designs for (**a**) single-stage and (**b**) two-stage continuous internal mixers [147]

mercially in a bewildering variety of designs and mixing capabilities [121]. Although extensively used in die-forming operations, for example, to manufacture polymer film, sheet or pipe, conventional single-screw extruders have only limited inherent mixing ability. Simple helical screw geometries subject the melt to only modest laminar shear strain and levels of shear stress generated are normally low [104].

This deficiency can be offset to some degree using distributive or dispersive mixing elements integrated in the overall screw design. (Figs. 23 and 24). These either create multiple disruptions to velocity profiles within the screw channel thereby randomizing the spacial location of fluid elements, or subject the material to localised shear stress by forced passage through narrow clearances [148]. With filled polymers they may assist in improving ultimate mixture quality prior to product formation, especially when using pre-dispersed filler in the form of master-batch compound and when additive levels are low, as in fibre-spinning or thin-film extrusion operations.

Static mixers offer an alternative route to enhancing mixture uniformity, through repeated division and combination of polymer passing through a tube containing motionless profiled elements (Fig. 23g). Mixers of this type closely control the applied shear strain developed and are positioned between the extruder and die [96].

However, for the majority of compounding requirements, conventional or modified single-screw extruder designs are inadequate, due to their inability to convey high levels of particulate filler and limited mixing intensity. As a consequence, special forms of compounding extruder have been designed to fulfill this function of which ko-kneaders and co-rotating twin-screw extruders have achieved most commercial prominence.

The *ko-kneader* has a unique mixing action in which, by using a special gear unit, a kneading screw can simultaneously rotate and reciprocate within the barrel. The screw flights are interrupted by three gaps per turn and the barrel has three rows of stationary kneading teeth projecting from the casing (Fig. 25). During operation, the twin motion of the screw causes the flights to pass between the teeth at each stroke forwards and backwards creating an interchange of material in both axial and radial directions. Furthermore, since the barrel pins wipe virtually all surfaces of the screw flights during this movement, high and controllable shear stresses can be applied. Many design and operational variations exist, including the length to diameter ratio of the machine, provision for downstream inclusion of additives into a pre-formed melt and a melt devolatilization capability. Process flexibility can be achieved by changing the screw geometry and kneading pins to achieve optimum mixing performance. Product from the ko-kneader is passed through a cross-head single-screw pump to generate pressure against a pelletizing die plate.

Despite the widespread use of this machine for compounding an extensive range of polymer-based formulations, only very limited analytical work has been reported on its operational performance. In one report, a modified flow analysis network method of simulation was used to describe flow of a Newtonian

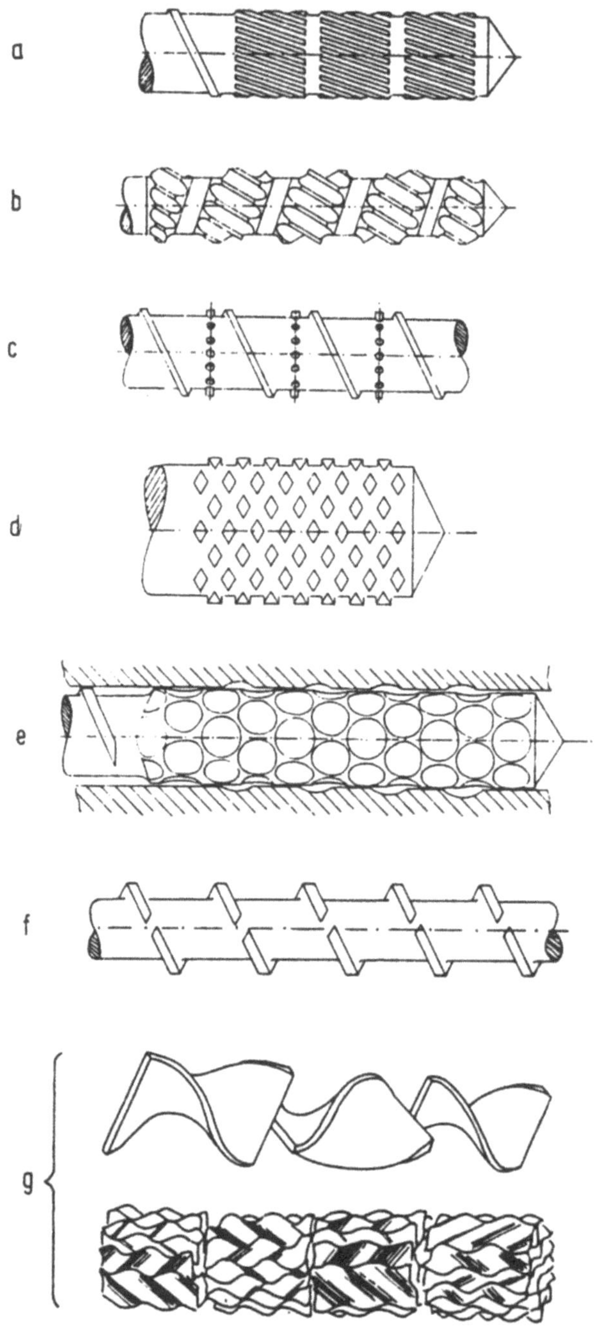

Fig. 23. Screw sections for enhancing distributive mixing in single-screw extruders (**a**) Dulmadge mixer, (**b**) Saxton mixer, (**c**) pin mixer, (**d**) pineapple mixer, (**e**) cavity transfer mixer, (**f**) slotted screw flight, (**g**) Kenics and Ross ISG static mixers [148]

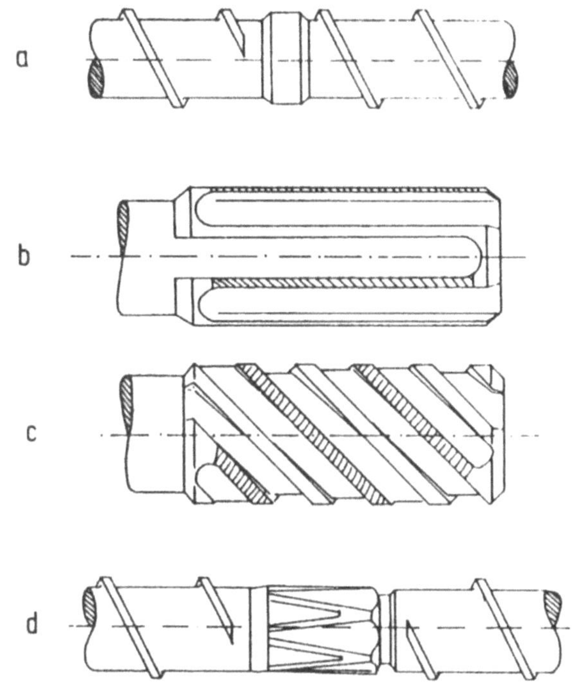

Fig. 24. Dispersive mixing sections for use in single-screw extruders (**a**) blister ring, (**b**) Union Carbide mixing section, (**c**) Egan mixing section, (**d**) Dray mixing section [148]

fluid under isothermal conditions. The model developed provided an estimate of local pressure distributions within the mixer and throughput characteristics [149].

Twin-screw extruders can be engineered to provide wide process flexibility through changes in screw and barrel configuration and operating procedure. As a consequence, a large number of design variants are available in which the screws may rotate in the same or opposite directions, intermesh to varying degrees or not at all, differ in geometry or flight profile, or contain elements which intensify polymer melting, mixing and conveying efficiency [150,151]. A detailed account of these various design permutations has been well documented, together with an appraisal of flow mechanisms and modeling studies undertaken [152].

However, much of the theoretical work published on twin-screw extruders provides only a partial insight into the operational behaviour of these machines, and fails to fully account for their complexity in multi-functional processing tasks involving conveying, melting, mixing and devolatization stages. Despite the extensive use of twin-screw extruders for the preparation of filled polymer composites, there are only a few detailed accounts on the effects of machine design and operation on compound quality.

(a)

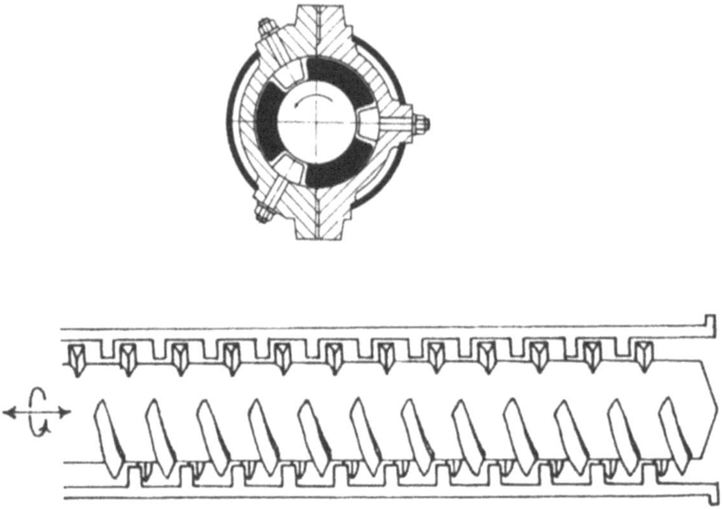

(b)

Fig. 25. Ko-kneader intensive compounding extruder [Courtesy Buss. AG] (**a**) Schematic diagrams of slotted screw and barrel pins, (**b**) ko-kneader with barrel in open position

The industrial use of twin-screw extruders for this purpose revolves extensively, but not exclusively, around intermeshing co-rotating variants. Closely intermeshing counter-rotating designs are widely used for profile extrusion of UPVC dry-blends since they permit close temperature control and exhibit a high conveying efficiency due to the positive displacement of material where the screws intermesh [150].

When melt compounding, effective material displacement is also of great significance, particularly when handling powdered feedstock and additives, and strongly influence product throughput. This aspect is of particular significance at very high levels of filler addition, for example, with ceramic formulations, where closely intermeshing screw elements with trapezoidal geometry have been proved to be highly effective [153]. Since mixture quality is of overriding importance, however, machinery must have sufficient flexibility to cater for the requirements of different material types, in relation to shear and thermal input. Optimum mixing almost always necessitates the introduction of special mixing elements, such as bilobal or trilobal kneading elements, slotted or interrupted screws and segmented discs into the screw design. Examples of these are shown in Fig. 26.

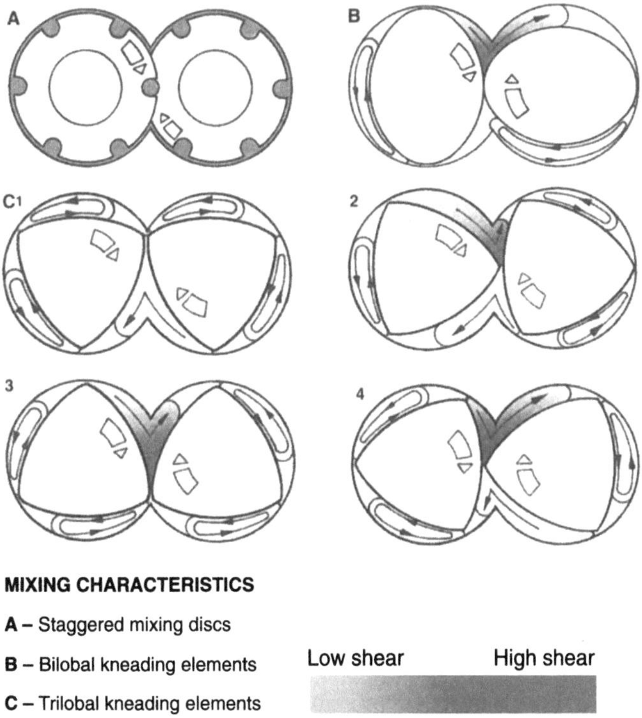

Fig. 26. Typical mixing elements for a co-rotating intermeshing twin-screw extruder

With kneading elements in particular, their mixing and conveying efficiency depends on the number of lobes, the width and number of elements, the clearance between the tip of the elements and barrel wall and degree of stagger in a forward or reverse helical direction [154]. The positioning of mixing elements within the screw profile also has important process implications. When located downstream towards the die they provide a distributive and dispersive mixing capability; however, when positioned nearer to the feed end they can contribute to polymer melting by providing a restriction to flow. However, it has been shown using mineral filled thermoplastics that dispersive mixing (measured as a mean volume particle diameter) is also greatly enhanced through the melting zone of a twin-screw extruder (Fig. 27). This has been attributed to the high shear stresses imposed on filler at the solid/melt interface.

Most conventional twin-screw compounding extruders permit interchangeable screw and barrel design, enabling the screw profile and/or barrel length to be modified, together with flexible assembly of the number and position of feedstock entry points and devolatilization locations. Figure 28 shows a typical screw and barrel assembly of a co-rotating intermeshing twin-screw extruder used for compounding filled thermoplastics and Fig. 29 the complete production line, with integration of necessary ancillary equipment for pellet manufacture. These requirements are discussed further below.

The modular screw and barrel concept has additional advantages, since regions where high wear occurs can be economically replaced without changing

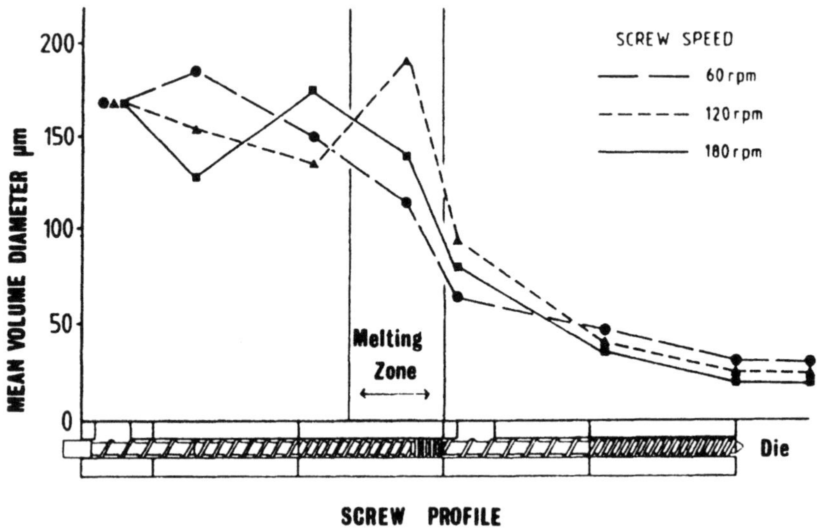

Fig. 27. Progressive dispersion of calcium carbonate in polypropylene within a co-rotating intermeshing twin-screw extruder. Filler dispersion is expressed in terms of a mean volume diameter determined by image analysis

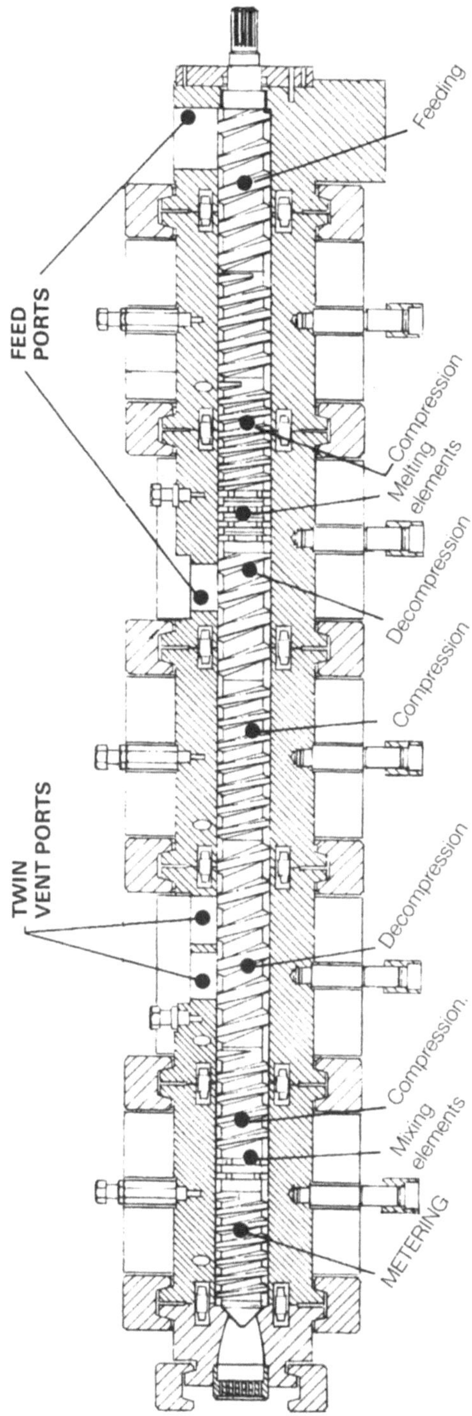

Fig. 28. Screw and barrel configuration of a co-rotating intermeshing twin-screw extruder for the preparation of filled thermoplastics compounds

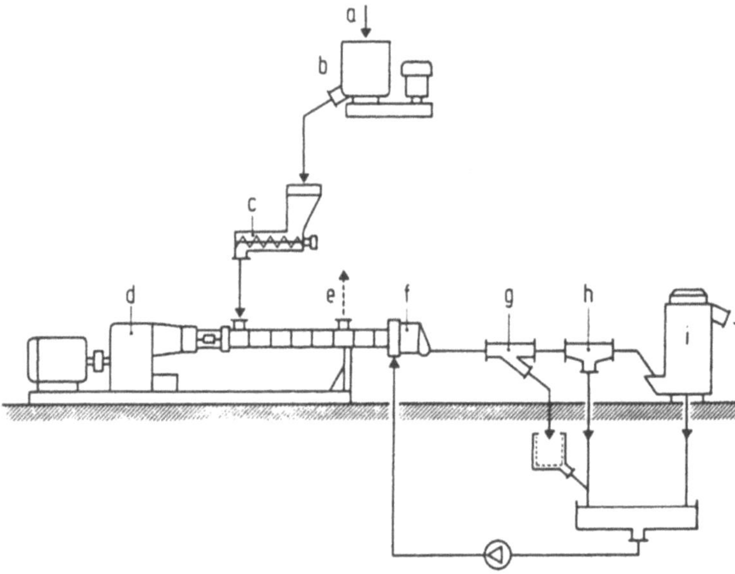

Fig. 29. Twin-screw extrusion compounding line for the preparation of mineral-filled thermoplastics showing pre-mixing, compounding and pelletizing stages [155] (**a**) Filler/polymer/additive feedstock, (**b**) high-speed pre-mixer, (**c**) dosing screw, (**d**) twin-screw compounder, (**e**) vacuum devolatilization, (**f**) water-cooled die-face cutter, (**g**) start-up diverter, (**h**) de-watering chute, (**i**) pellet dryer, (**k**) bagging

the rest of the screw. This will inevitably occur when combining polymers with abrasive inorganic fillers, particularly at high addition levels. Wear resistant inserts can be used to protect barrel sections [155].

Although the function of the vast majority of twin-screw extrusion compounding lines processing particulate fillers and thermoplastics generate pelletised compound, integration of compounding and end-fabrication stages can lead to reduced manufacturing costs, and improved product quality. For example, in-line compounding of filled polymers into profile, such as film, sheet and pipe, can offer economic benefits in production over two-stage processing, which requires the compound to be prepared in pellet form then re-extruded [156,157]. Importantly, with heat and shear sensitive fillers, only one processing cycle is experienced. However, capital equipment costs are significantly higher than conventional profile manufacturing lines based on single-screw extruders. Furthermore, incorporation of gear pumps between compounding extruder and die are generally recommended to stabilise pressure fluctuation, which is inevitable from the twin-screw extruder output.

The concept of single stage product manufacture may also be applied to injection moulding technology as shown in Fig. 30 [158]. Compound preparation

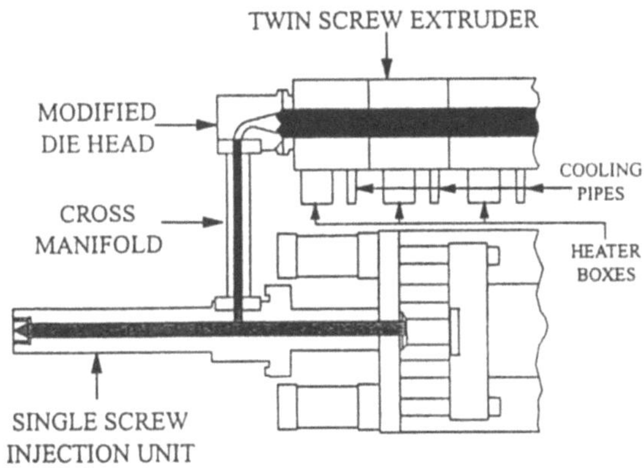

Fig. 30. Design principles of direct compounding injection moulding machine for processing highly filled polymer compositions

is carried out on a co-rotating twin-screw compounding extruder, which subsequently injects the well-blended mixture into the mould cavity. This technology is particularly well suited for compounding and moulding very highly filled ceramic or metal injection moulding formulations into complex artifacts [159] but can readily be applied to many other particulate-filled polymer compositions.

Integrated polymer compounding technology has also been developed for the preparation of polymer composites containing low-cost reinforcing additives derived from waste products, including natural fibre reinforcements obtained from agricultural sources and fibre-reinforced thermoset scrap [160]. The method is based on modified twin-screw extrusion technology requiring assembly of well-defined process steps which prepare and condition the additive phase *prior* to its incorporation into the polymeric matrix, which may be thermoplastic, thermosetting or elastomeric. This concept is the inverse of conventional extrusion compounding procedures, where filler is added together with, or subsequent to, the polymeric feedstock (Fig. 31).

Suitably comminuted thermoset scrap containing reinforcing fibres, for example, can be regarded as a functional filler which has been shown to significantly enhance the mechanical properties of the base polymer and with phenolic based compositions may confer additional benefits such as improved fire performance to the material [161]. The method involves an initial size reduction step which controls the physical size and morphology of the filler particles. This is combined with a treatment stage which allows the surface chemistry of the filler to be modified to promote bonding with the matrix. Introduction of treatment in this way overcomes the difficulties of applying conventional treatment technology to irregular recyclate particles, which may also undergo further breakdown during subsequent mixing operations, thereby exposing uncoated additive surfaces.

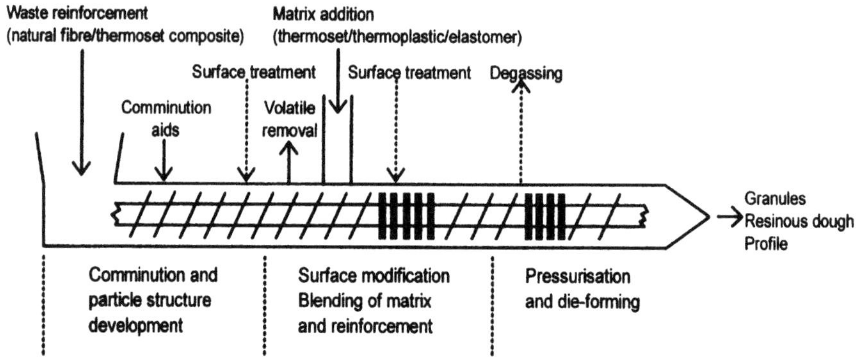

Fig. 31. Integrated twin-screw extrusion compounding technology for the preparation of polymer composites filled with fibrous recyclate additives

Molten or resinous polymer is then introduced into the integrated process unit and combined with the treated filler. Other stages, including addition of further treatment, devolatilization, pressurization and die-forming may also be required depending on the nature of the composition and its intended application.

4.3.3 *Ancillary Operations*

It will be apparent from the preceding discussion that most continuous compounding operations for the preparation of filled thermoplastics require integration of a variety of ancillary operations designed to accurately meter polymer feedstock and additives into the compounder and, if necessary, filter the homogenised composition whilst in a melt state, to remove extraneous matter or undispersed particles which might impair mechanical properties of the filled composite. Converting the melt into a solidified pellet of uniform geometry also requires specialist equipment which can differ in design and complexity.

Due to the variable flow characteristics of powdered fillers, and the range of addition levels often demanded (for example, from 20 to 80% by weight), it is not possible to design a unified feed system for all materials [155]. Considerations include the operational design and accuracy required from the feeders, the location of filler addition (within solid or melt zones, or a combination of both) and the need to utilise a stuffer screw. A variety of feeding options are illustrated in Fig. 32.

Commercial feeders commonly operate using volumetric or gravimetric principles. The former type delivers material using single or twin screws and can be very sensitive to fluctuations in material bulk density . Gravimetric feeders give a much higher degree of accuracy by delivering materials on a mass basis. For example, with loss-in-weight feeders, a storage hopper with bulk material

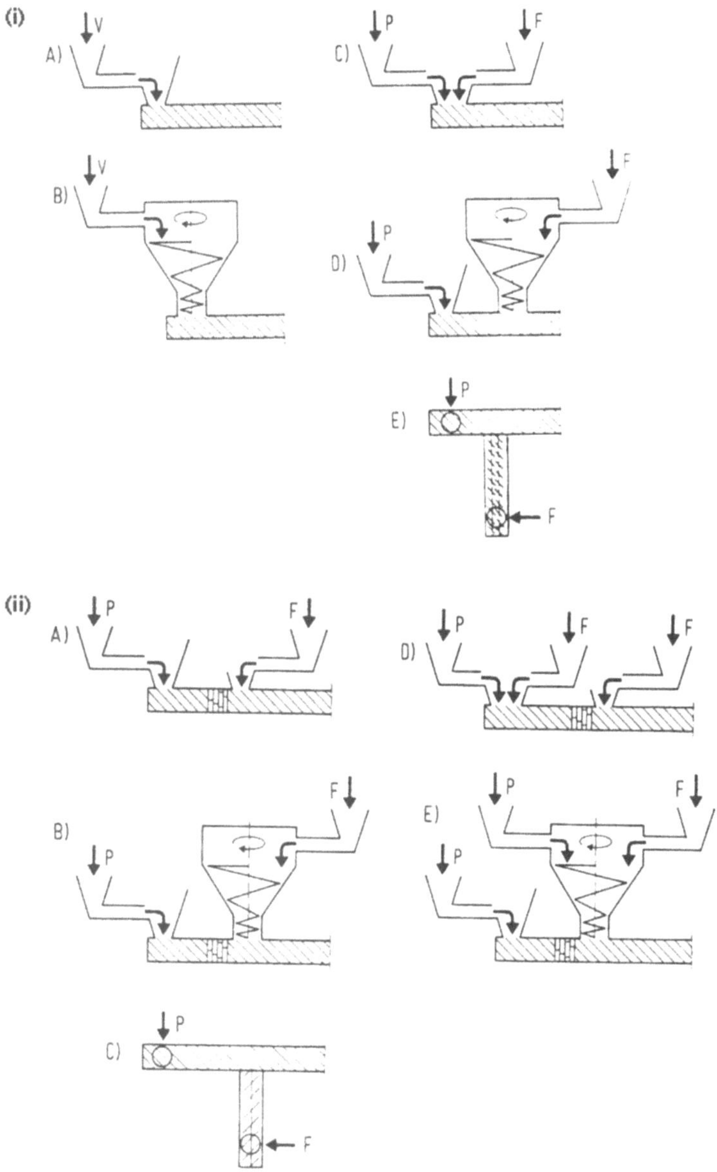

Fig. 32. Options for introducing filler and polymer feedstock into continuous compounding extruders. **(i)** Filler addition to the solids zone – (**a**) premix free fall, with partial degassing; (**b**) premix introduced through stuffer screw; (**c**) separate addition by free-fall of filler and polymer; (**d**) separate dosing of filler and polymer using single-screw metering unit; (**e**) separate feeding of filler and polymer using side mounted twin-screw feed unit. **(ii)** Downstream addition of filler into the polymer melt – (**a**) by free fall; (**b**) using a single-screw feed unit; (**c**) with side-mounted twin-screw feeder; (**d**) split-feeding of filler in both solids and melt zones; (**e**) split-feeding of polymer in solids and melt zones. (F: filler; P: polymer; V: premix of filler and polymer) [155]

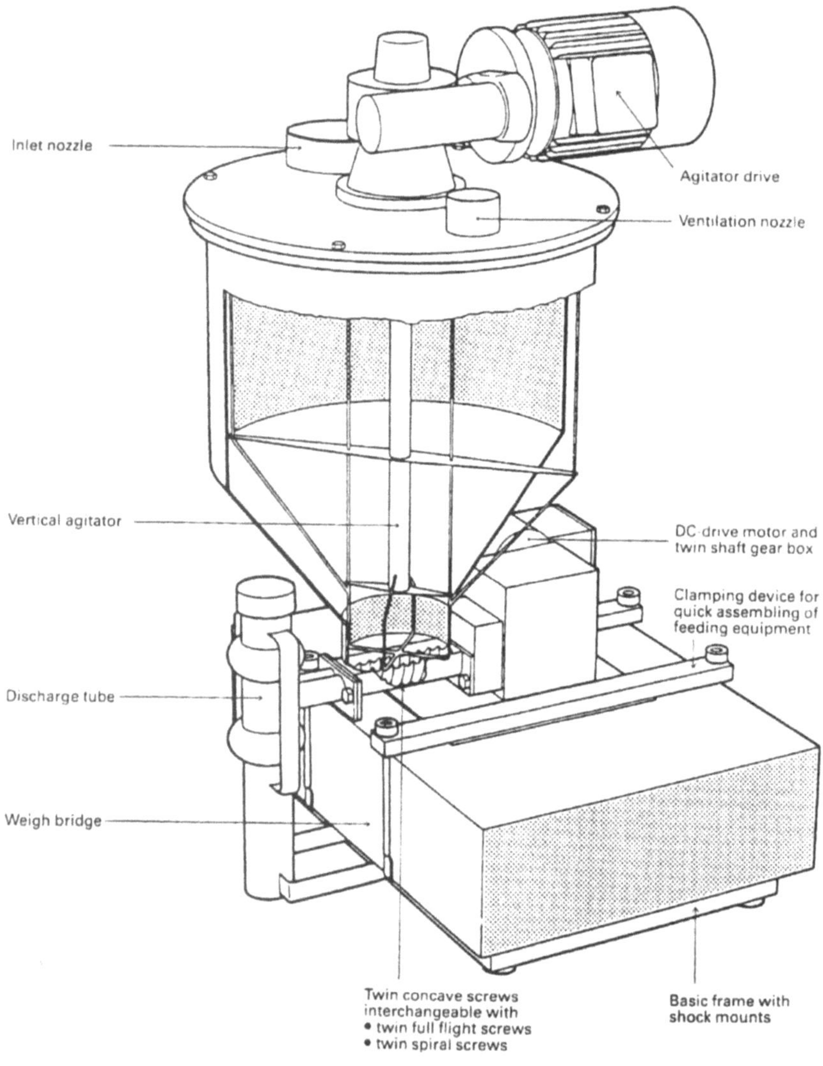

Fig. 33. Constructional features of loss in weight gravimetric feeder.

and delivery screw feeder are placed on a weigh bridge. (Fig. 33). During feeding a microprocessor controller compares the weight reduction per unit time against a pre-determined set-point value. Deviations are corrected by adjusting the drive speed of the feeder. During the short period when the hopper requires refilling, the controller regulates delivery rate using memorised feeding values. This approach can give an accuracy of ±0.25% to 1%.

Melt filtration systems are commonly employed in pigment master-batch production and in situations where the presence of defects in the compound may have a critical effect on its subsequent processing or properties. This is vitally important, for example, in fibre-spinning operations involving extrusion of polyester or polyamide through fine spinneret plates [162], and in minimizing breakdown of polymer cable insulation subjected to electrical stress [163].

Model investigations undertaken using high-density polyethylene pipe dosed with particulate aluminum flaws of known size showed that resistance to stress rupture increased significantly after removal of larger stress raising particles by melt filtration [164].

Alternative routes exist for the preparation of compound granules. With continuous compounding plant melt is extruded through a strand die, then cooled and pelletised using a rotating cutter. However, when the compound melt strength is low, the cooled strands are brittle or throughputs are high, hot die-face cutting with an eccentrically mounted rotating knife is the preferred option. Pellets are cooled by water or air, which also carries them away from the die face. Additional facilities for de-watering and drying of granules may also be required (Fig. 29).

5 Structure Development in Melt Processed Particulate-Filled Polymer Composites

The properties of particulate-filled thermoplastics depend on many factors, including the nature of the filler, the quantity added, surface interaction with the matrix and the quality of mixing achieved. However, several filler types are amenable to orientation or structuring during processing, which may further influence or enhance their performance. For example, these effects may result from flow through dies or into mould cavities and during melt deformation in thermoforming or blow moulding operations.

Investigations undertaken using polyethylene containing carbon black, calcium carbonate or titanium dioxide filler, showed that filler levels, in excess of 5% by volume, suppressed the development of vortices, usually observed at the entrance region of 180° entrance angle dies [165,166]. Similar behaviour was obtained in diverging entrance dies and those with offset or multiple hole entries. This behaviour was attributed to differences in the rheological properties of the melt due to the presence of filler, possibly associated with changes to elongational viscosity from an increasing rather than decreasing function of stretch rate.

The mechanical properties of thermoplastics containing short fibre reinforcement are strongly influenced by the direction of fibre alignment relative to the applied stress. Injection-moulded artifacts, for example, can exhibit complex fibre orientation patterns dependent on the mould geometry and conditions experienced by the melt during filling of the mould cavity [30]. Plate-like filler particles, such as talc, mica, glass, or aluminum flakes are also anisotropic, but to a lesser extent than high aspect ratio fibres and may also orientate in thermoplastics melts during processing. This phenomenon has been considered using a variety of techniques including microradiography, scanning electron microscopy and wide angle X-ray diffraction (WAXD) [167].

Using thermoplastics containing mica, talc and Kevlar fibres, WAXD pole figures were obtained after subjecting the materials to different processing histories [168]. Operations were chosen to achieve (i) flows with uniaxial symmetry about a reference direction (capillary die extrusion, melt spinning), (ii) flows with planar tendencies and a preferred direction (slit die extrusion, compression moulding and cold rolling), and (iii) flows with a relatively weak relationship to reference symmetries (injection moulding). Generally, it was found that plate-like particles orient so that their major surfaces lie parallel to the metal walls of the die or mould prescribing the flow. With injection-moulded specimens in the intermediate and skin layers, high levels of mica and talc flake orientation were observed in machine and transverse directions, but this was lower in the centre of the moulded part. Similarly, talc flakes showed preferred alignment parallel to the surface of blow-moulded and thermoformed polyolefin parts [169,170].

With mixed aramid fibre/particulate flake compositions in polycarbonate, however, the previously mentioned processing operations significantly reduced the levels of fibre and flake orientation relative to corresponding observations with single particle compounds [167]. For example, in sheet extrusion and injection moulding, the flakes oriented perpendicular to the fibres losing their biaxial orientation parallel to the surrounding machine surfaces.

Using X-ray diffraction and scanning electron microscopy it was shown for talc-filled polypropylene compounds that in the skin region of injection mouldings the long axis of the talc platelets aligned parallel to the moulding surface, whereas through the moulding thickness orientation was more random [171]. In this work, 6-mm thick plaques were used, whereas with the work involving talc and mica cited above, only 3-mm thick specimens were analyzed.

Unlike conventional injection moulding technology where molecular and particle orientation effects are determined primarily by the flow conditions experienced by the melt as it flows into the cavity and cools under a static pressure, shear controlled orientation injection moulding (SCORIM) can apply a macroscopic shear to the melt as it is cooling, thereby influencing microstructure [172,173]. This is achieved by splitting the usual single feed from the injection moulder into a plurality of live feeds, which can independently control pressure to the cavity, using auxiliary packing pistons (Fig. 34). The position of the feeds and sequencing of these pistons after filling the mould cavity influences the structure of the solidifying melt at the solid/melt interface generating high levels of fibre orientation in short fibre-reinforced thermoplastics materials.

The application of this technique to talc-reinforced polypropylene has shown that the microstructure of platelets and the resulting physical properties of the moulded composites are markedly affected [171)]. With two live feeds located at either end of the mould cavity, the talc platelets exhibited strong talc platelet alignment throughout the thickness of the moulding in the direction of the ap-

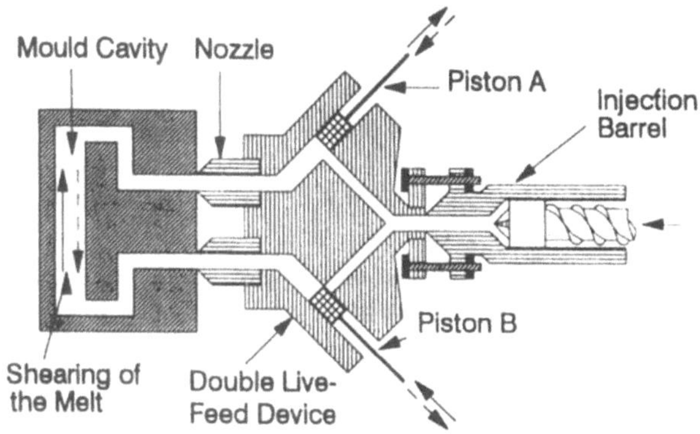

Fig. 34. Principles of SCORIM multiple live feed injection moulding technology. To impart shear on the melt-solid interface as this progresses from the surface to the centre of the cavity, pistons A and B are operated 180° out of phase

plied shear and gave a greater stiffness in this direction relative to conventionally processed mouldings. Four live feeds positioned symmetrically on faces of a square plaque moulding with sequential shearing between opposite pairs of gates gave rise to a biaxial talc orientation relative to the plaque surface and more isotropic, although slightly reduced, stiffness compared to the double live-feed variant.

In addition to the nature of particulate platelet orientation induced during injection moulding, the associated consequences on molecular orientation and crystalline order of the host thermoplastic matrix have also been reported with particular regard to various flake-filled polypropylenes [174], together with an attempt to interrelate these higher order structural parameters with physical properties of the composites [175].

Structural analysis of injection-moulded parts made from polystyrene reinforced with spherical glass beads has demonstrated that significant segregation of the particulate phase occurs as a result of the flow experienced during filling of the mould cavity [176]. Mouldings examined revealed a region of particles at the cavity mid-plane surrounded by a particle-free zone. This was frequently surrounded by monolayers attached to the moulding surfaces. In addition, there was a gradual accumulation of particles towards the advancing free surface, observed during mould filling, and an increase in particle concentration with distance from the gate. The reported phenomena were shown to be influenced by the particle size and by the geometry and location of the injection gate. In mouldings with inserts, the formation of particle-rich weld lines were observed. The results were explained in terms of the combined effects of convection, fountain flow and lateral particle migration occurring during mould filling.

Using high levels (90% by weight) of magnetic strontium ferrite filler particles in a polyamide 6 matrix, procedures have been described for structuring the filler particles within the cavity of an injection moulding process [177]. Moulded products with high magnetic intensity can be achieved by generating magnetic fields within the mould cavity using externally mounted coils (Fig. 35) thereby influencing the particle orientation before melt cooling. Optimization of the process requires particular consideration of the coiling system, mould construction and pole geometry.

The electrical conductivity of thermoplastics containing carbon black filler can be strongly influenced by the method of processing and its effect on the microstructure of the composite [178,179]. For example, variations in electrical resistivity were measured within polyethylene injection mouldings containing 20% by weight of carbon black [180]. Resistivity was found to increase with orientation, with values in the skin and regions of high shear being much higher than in the core. In the case of surface resistivity measurements, the difference between the biaxially oriented skin and virtually unoriented core amounted to six decades for LDPE, however HDPE conductivity was slightly less sensitive to orientation. There was no direct relationship between the resistivity values and molecular orientation measured by thermal shrinkage, of 50 μm thick microtome slices taken at increasing distances from the surface. The effects observed were interpreted in terms of disruption of the conducting carbon black network during filling of the mould cavity. On annealing, the conductive network was allowed to re-form and the resistivity decreased to its normal value for unoriented polymer.

Bayer et al. have examined the electrical properties of carbon black/polyethylene composites processed using elongational flow injection moulding [181]. Mould geometry was optimised in the form of double-armed bars so as to give enhanced mechanical properties combined with a high degree of electrical homogeneity. A high molecular weight linear polyethylene (MW 450,000) was precompounded with filler concentrations from 1 to 7.5% by volume.

In marked contrast to conventional injection moulding, where orientation effects normally depress conductivity, in this investigation the injection-moulded composite material yielded not only a lower percolation threshold than compression moulded samples, but in the injection direction, also gave conductivity values two to three orders of magnitude higher than the latter.

In the radial direction at low carbon black concentration (4%), conductivity data exactly paralleled birefringence changes, with peak values in the high shear zones near to the surface with a minimum at the core. At higher filler concentrations, however, the conductivity profile exhibited a broad homogeneous maximum in the centre with a conductivity level of 10^{-1} ohm^{-1} cm^{-1}. This was attributed to an increase in melt viscosity leading to broadening of the shearing zones within the mould cavity, overlapping in the centre, resulting in the formation of a uniform conductive stiff inner cylinder, several millimeters wide, homogeneously extending along the full length of the injected material. Microstructurally, the material showed the existence of oriented polyethylene shish-kebabs exhibiting segregated axial channels of conducting carbon black.

A subsequent investigation considered further the effects on conductivity of these induced microstructural features, but with specific attention to changes in

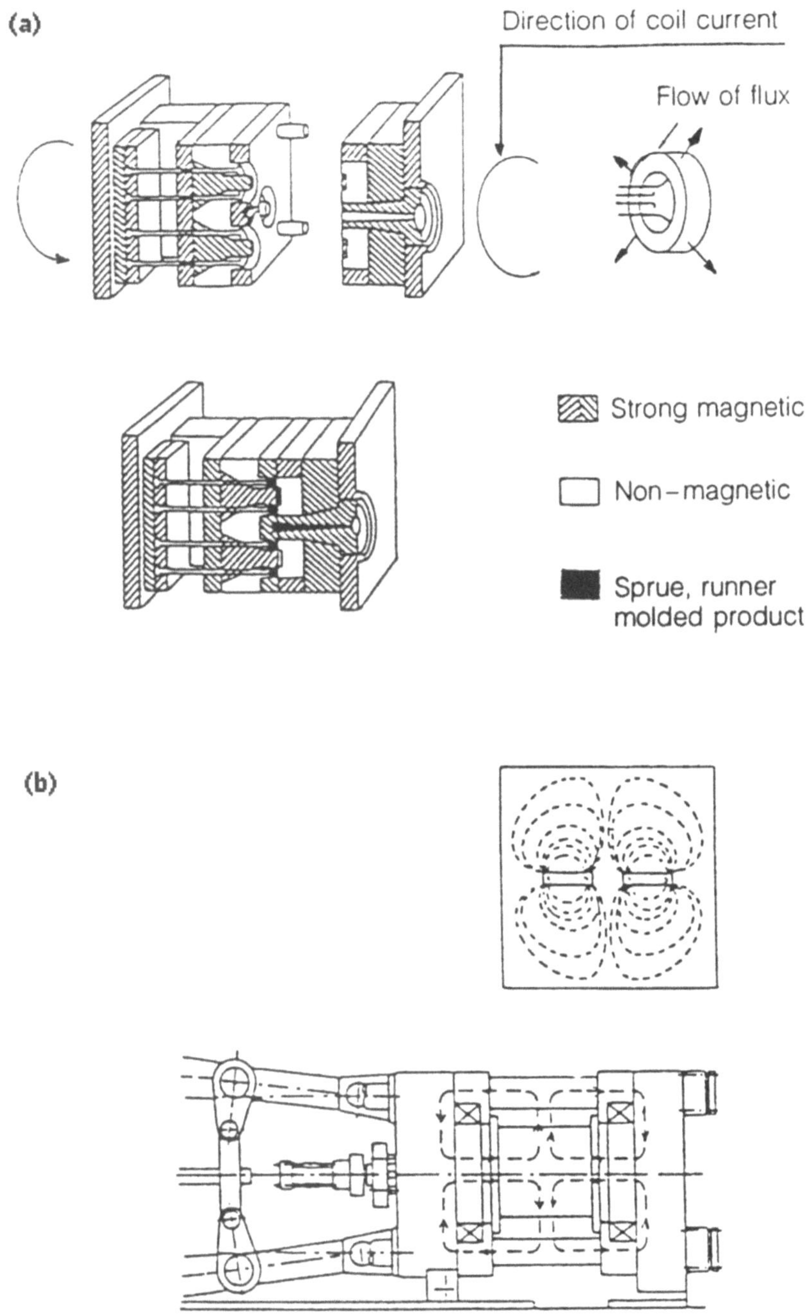

Fig. 35. In-mould magnetization injection moulding technology for filled thermoplastics: (**a**) constructional features of mould; (**b**) magnetic flux generation in the radial direction [177]

the polymer molecular weight [182]. Thus polymer composites containing carbon black were prepared by elongational flow injection moulding, and compression moulding, using linear HDPE polyethylene with lower molecular weights (M_η) ranging from 51,000 to 248,000.

The conductivity results show a strong dependency on polymer molecular weight and are summarised in Fig. 36 for composites containing 4% by volume of carbon black. Thus, at low and intermediate molecular weights, the conductivity decreased after injection moulding and material orientation (see I). This effect coincides with findings from other works, such as Kubát mentioned earlier. The observed decrease in conductivity relative to compression moulded materials was explained in terms of increased orientation of the anisotropic carbon black aggregates. However, as polymer molecular weight increases, the shear and elongational stress contribution in the matrix during injection moulding in-

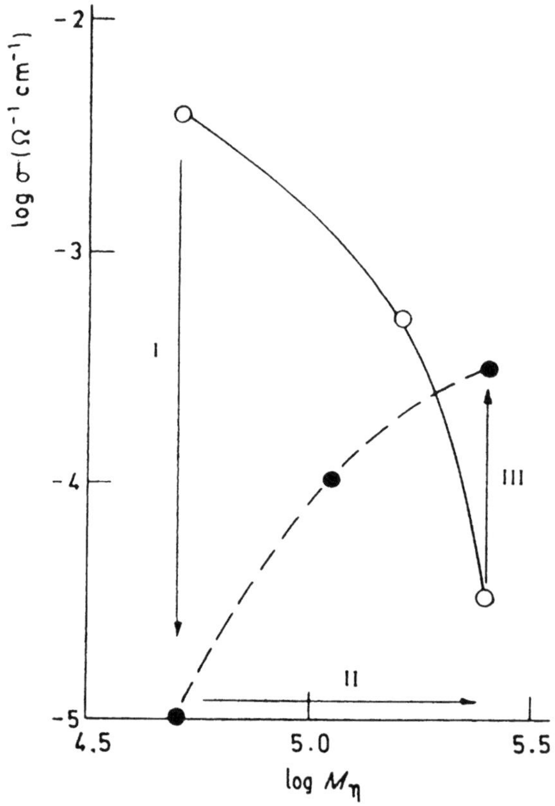

Fig. 36. Variation in electrical conductivity (σ) with molecular weight for polyethylene composites filled with 4% by volume carbon black, demonstrating the effects of: orientation (I), degradation (II) and flow-induced segregation of carbon black aggregates (III). (●) injection moulded (○) compression moulded (unoriented) [181]

creases (c.f. earlier discussion on dispersion of carbon black agglomerates) leading to a breakdown of these aggregates and, as a consequence, reduced anisotropy (see II). This has the effect of decreasing conductivity in isotropic compression moulded samples, due to a less well-defined percolation conductive filler network. However, for injection-moulded materials, particularly using the highest molecular weight matrix, even though the filler is highly degraded, flow-induced segregation of these additive particles becomes dominant, generating highly conductive channels, and a value of conductivity significantly higher than that for unoriented material with a nearly homogeneous particle distribution (see III in Fig. 36).

6
References

1. Malkin, A.Y. (1990) Advances in Polymer Science 96: 69
2. Carreau, P.J., Lavoie, P-A. (1996) Macromol. Symp. 108: 111
3. Einstein, A. (1906) Ann. Phys 19:289, (1911) 34: 591
4. Nielsen, L.E. (1977) Polymer Rheology, Dekker
5. Rutgers, R. (1962) Rheol. Acta 2: 305
6. Mooney, M. (1951) J. Colloid Science 6: 162
7. Maron, S.H., Pierce, P.E. (1956) J. Coll. Sci. 11: 80
8. Krieger, I.M., Dougherty, T.J. (1959) Trans. Soc. Rheol. 3: 137
9. Lewis, T.B., Nielsen, L.E. (1968) Trans. Soc. Rheol. 12: 421
10. Cross, M.M. (1970) J. Colloid Interface Sci. 33: 30
11. Cross, M.M. (1965) J. Colloid Sci. 20: 417
12. Mewis, J., Frith, J.W., Strivens, T.A., Russel, W.B. (1989) AIChE Journal 35 (3): 415
13. Poslinski, A.J., Ryan, M.E., Gupta, R.K., Seshadri, S.G., Frechette, F.J. (1988) J. Rheol. 32 (7): 703, 751
14. Casson, N., Mill, C. (1959) Rheology of disperse systems. Pergamon Press, New York
15. Brydson, J.A. (1981) Flow properties of polymer melts. George Godwin, London
16. Piau, J.M., Agassant, J-F. (1996) Rheology for polymer melt processing. Elsevier, Amsterdam
17. Han, C.D. (1981) In: Multiphase flow in polymer processing. Academic Press, chap 3
18. Minagawa, N., White, J.L. (1976) J. Appl. Polym. Sci. 20: 501
19. Lobe, V.M., White, J.L. (1979) Polym. Eng. Sci. 19:617
20. Sothern, G.R., Hodd, K.A. (1981) Proceedings from Filplas '81 Conference, London British Plastics Federation
21. Quackenbush, C.L., French, K., Neil, J.T. (1982) Proc. Ceram. Eng. Soc. 3: 20
22. Edirisinghe, M.J., Evans, J.R.G. (1987) J. Mat. Sci. 22: 269
23. Edirisinghe, M.J., Evans, J.R.G. (1987) Br. Ceram. Trans. J. 86: 18
24. Stedman, S.J., Evans, J.R.G., Woodthorpe, J. (1990) J. Mat. Sci. 25: 1833
25. Issitt, D.S., James, P.J. (1986) Powder Metall. 29: 259
26. Song, J.H., Evans, J.R.G. (1996) J. Rheol. 40 (1): 131
27. Barnes, H.A. (1989) J. Rheol. 33: 329
28. Chong, J.S., Christiansen, E.B., Baer, A.D. (1971) J. Appl. Polym. Sci. 15: 2007
29. White, J.L., Crowder, J.W. (1974) J. Appl. Polym. Sci. 18: 1013
30. Folkes, M.J. (1982) Short fibre reinforced thermoplastics. Research Studies Press, Chichester, UK
31. Utracki, L.A. (1986) Polymer Composites 7 (5): 274
32. Utracki, L.A., Favis, B.D., Fisa, B. (1984) Polymer Composites 5 (4): 277
33. Fisa, B., Utracki, L.A. (1984) Polymer Composites 5 (1): 36

34. Utracki, L.A. (1984) Rubber Chemistry and Technology 57: 507
35. Suh, C.H., White, J.L. (1996) J. Non-Newtonian Fluid Mech. 62: 175
36. Hornsby, P.R., Mthupha, A. (1994) J. Mat. Sci. 29: 5293
37. Kim, K.J., White, J.L. (1996) J. Non-Newtonian Fluid Mech. 66: 257
38. Rochette, A., Choplin, L., Tanguy, P.A. (1988) Polymer Composites 9 (6): 419
39. Stamhuis, J.E., Loppe, J.P. (1982) Rheol. Acta 21: 103
40. Suetsugu, Y., White, J.L. (1983) J. Appl. Poly. Sci. 28: 1481
41. Han, C.D., Sandford, C., Yoo, H.J. (1978) Polym. Eng. Sci. 18 (11): 849
42. Rothon, R.N. (1995) In: Rothon, R.N. (ed.) Particulate-filled polymer composites. Longman, Harlow, chap 4
43. Malik, T.M., Carreau, P.J., Grmela, M., Dufresne, A. (1988) Polymer Composites 9 (6): 412
44. Herschel, W.H., Buckley, R. (1926) Proc. Am. Soc. Test. Mater. 26: 621
45. Herschel, W.H., Buckley, R. (1926) Kolloid-Z. 39: 291
46. Kambe, H., Takano, M. (1965) Proc. Int. Congr. Rheol. 4th Part 3 557 Wiley-Interscience, New York
47. Lin, L., Masuda, T. (1990) Poly. Eng. Sci. 30 (14): 841
48. Gahleitner, M., Bernreitner, K., Neissl, W. (1994) J. Appl. Polym. Sci. 53: 283
49. Gahleitner, M., Bernreitner, K., Knogler, B., Neissl, W. (1996) Macromol. Symp. 108: 127
50. Osanaiye, G.J., Leonov. A.I., White, J.L. (1993) J. Non-Newtonian Fluid Mech. 49: 87
51. Lakdawala, K., Salovey, R. (1987) Poly. Eng. Sci. 27 (14): 1035
52. Donnet, J-B., Bansal, R.C., Wang, M-J. (1993) Carbon black: science and technology, 2nd edn. Marcel Dekker, New York
53. Noguchi, T., Nagai, T., Seto, J. (1986) J. Appl. Poly. Sci. 31: 1913
54. Aranguren, M.I., Mora, E., DeGroot, J.V., Macosko, C.W. (1992) J. Rheol 36 (6): 1165
55. White, J.L., Tanaka, H. (1981) J. Appl. Polym. Sci. 26: 579
56. Lobe, V.M., White, J.L. (1979) Polym. Eng. Sci. 19: 617
57. Chan, Y., White, J.L., Oyanagi, Y. (1978) J. Rheol. 22: 507
58. Han, C.D., Kim, Y.W. (1974) J. Appl. Polym. Sci. 18: 2589
59. Trouton, F.T. (1906) Proc. Roy. Soc. A-77: 426
60. Utracki, L.A., Lara, J. (1984) Polymer Composites 5 (1): 44
61. Greener, J. (1995) Elongational Flow in Ceramics Processing. PhD thesis, Brunel University, UK
62. Tanaka, H., White, J.L. (1980) Polym. Eng. Sci. 20: 949
63. Dreval, V.E., Borisenkova, E.K. (1993) Rheol. Acta. 32: 337
64. Carr, W. (1978) J. Oil Col. Chem. Assoc. 61: 397
65. Carr, W. (1982) J. Oil Col. Chem. Assoc. 65: 373
66. Crowl, V.T. (1972) J. Oil Col. Chem. Assoc. 55: 388
67. Varley, D.M., Bower, H.H. (1979) J. Oil Col. Chem. Assoc. 62: 401
68. Lee, Y-J., Manas-Zloczower, I., Feke, D.L. (1995) Polymer Engineering and Science 35 (12): 1037
69. Ahmed, M. (1979) In: Coloring of plastics: theory and practice. Van Nostrand Rheinhold, New York, chap 6
70. Smith, M.J. (1973) J. Oil Col. Chem. Assoc. 56: 165
71. Rwei, S-P., Manas-Zloczower, I., Feke, D.L. (1992), Poly. Eng. Sci. 32 (2): 130
72. Bigg, D.M. (1984) Advances in Polymer Technology 4 (3/4): 255
73. Van Drumpt, J.D. (1988) Plastics Compounding March/April: 37
74. Riley, A.M., Paynter, C.D., McGenity, P.M., Adams, J.M. (1990) Plastics Rubber Process Appl. 14: 85
75. Suetsugu, Y. (1990) Intern. Polymer Processing 5 (3): 184
76. Nakajima, N., Shieh, W.J., Wang, Z.G. (1991) Intern. Polymer Processing 6 (4): 290
77. Wagenknekt, U. (1996), Proceedings from Loughborough Fillers Symposium II., Loughborough University, UK

78. Hornsby, P.R., Hinrichsen, E., Tarverdi, K. (1986) J. Mat. Sci. 32: 443
79. Edirisinghe, M.J., Evans, J.R.G. (1986) Proc. Brit. Ceram. Soc. 38: 67
80. Kolarik, J., Jancar, J. (1992) Polymer 33 (23): 4961
81. Wang, J., Tung, J.F., Ahmed Fuad, M.Y., Hornsby, P.R. (1996) J. Appl. Poly. Sci. 60: 1425
82. Sumita, M., Sakata, K., Asai, S., Miyasaka, K., Nakagawa, H. (1991) Polymer Bulletin 25: 265
83. Manas-Zloczower, I., Nir, A., Tadmor, Z (1984) Rubber Chemistry & Technology 57: 583
84. Rumpf, H., Schubert, H. (1978) In: Onada, G.Y., Hench, L.L. (eds) Ceramic processing before firing. Wiley-Interscience, chap 27
85. Cheng, D.C-H. (1968) Chemical Engineering Science 23: 1405
86. Schubert, H., Herrmann. W., Rumpf, H. (1975) Powder Technology 11: 121
87. Newitt, D.M., Conway-Jones, J.M. (1958) Trans. Instn. Chem. Engrs. 36: 422
88. Smith, M.J. (1973) J. Oil Col. Chem. Assoc. 56: 126
89. Smith, M.J. (1973) J. Oil Col. Chem. Assoc. 56: 155
90. Niesz, D.E., Bennett, R.B. (1978) In: Onada, G.Y., Hench, L.L. (eds) Ceramic processing before firing. Wiley-Interscience, chap7
91. Horwatt, S.W., Rwei, S-P.,, Manas-Zloczower, I. (1989) Rubber Chemistry and Technology 62: 928
92. Drakopoulou, A, Feke, D.L. (1993) Rubber Chemistry and Technology 67: 17
93. Smith, M.J. (1974), J. Oil Col. Chem. Assoc. 57: 36
94. Ess, J.W., Hornsby, P.R. (1987) Plastics & Rubber Processing and Applications 8: 147
95. Tadmor, Z., Gogos, C.G. (1979) Principles of Polymer Processing. Wiley, New York
96. Middleman, S. (1977) In: Fundamentals of polymer processing. McGraw-Hill, New York, chap 12
97. Danckwerts, P.V. (1952) Appl. Sci. Res. Section A 3: 279
98. Ess, J.W., Hornsby, P.R., Lin, S.Y., Bevis, M.J. (1984) Plastics and Rubber Processing and Applications 4: 7
99. Ess, J.W., Hornsby, P.R. (1986) Polymer Testing 6:205.
100. Mohr, W.D., Saxton, R.L., Jepson, C.H. (1957) Industrial and Engineering Chemistry 49 (11): 1855
101. Spencer, R.S., Wiley, R.M. (1951) J. Colloid Science 6: 133
102. Danckwerts, P.V. (1953) Chemical Engineering Science 2 (1): 1
103. Hornsby, P.R., Singh, D.P., Sothern, G.R. (1985) Polymer Testing 5: 77
104. Tadmor, Z., Klein, I. (1978) Engineering principles of plasticating extrusion. Krieger Publishing Co., New York
105. Medalia, A.I. (1970) J. Colloid Interface Science 32: 115
106. McKelvey, J.M. (1962) Polymer processing. Wiley-Interscience, New York
107. Dizon, E.S., Micek, E.J., Scott, C.E. (1975) Rubber Chem. Technol. 48: 339
108. Tadmor, Z. (1976) Ind. Eng. Chem. Fundam. 15: 346
109. Manas-Zloczower, I., Feke, D.L. (1989) Intern. Polymer Processing 4: 3
110. Coran, A.Y., Ignatz-Hoover, F., Smakula, P.C. (1994) Rubber Chemistry & Technology 67: 237
111. Chohan, R.K., David, B., Nir, A., Tadmor, Z. (1987) Intern. Polymer Processing 2: 13
112. Hess, W.M., Ban, L.L., McDonald, G.C. (1969) Rubber Chem. Technol. 42: 1209
113. Voet, A., Aboytes, P., Marsh, P.A. (1969) Rubber Age 101: 78
114. Rwei, S-P., Horwatt, S.W., Manas-Zloczower, I., Feke, D.L. (1991) Intern. Polymer Processing 6: 98
115. Rwei, S-P., Manas-Zloczower, I., Feke, D.L. (1991) Poly. Eng. Sci. 31 (8): 558
116. Rwei, S-P., Horwatt, S.W., Manas-Zloczower, I., Feke, D.L. (1991) Intern. Polymer Processing 6: 2
117. Horwatt, S.W., Manas-Zloczower, I. (1992) Rubber & Chemistry Technology 65: 805
118. Shiga, S., Furuta, M. (1985) Rubber Chem. Technol. 58: 1
119. Lee, Y-J., Manas-Zloczower, I., Feke, D.L. (1993) Chem. Eng. Science 48: 3363

120. Bohin, F., Manas-Zloczower, I., Feke, D.L. (1994) Rubber & Chemistry Technology 67: 602
121. Rauwendaal, C. (1986) Polymer extrusion. Hanser, Munich
122. Biesenberger, J.A. (ed.) (1983) Devolatilization of polymers. Hanser, Munich
123. Matthews, G. (1982) Polymer mixing technology. Applied Science, London
124. Matthews, G. (1972) Vinyl and allied polymers. Iliffe, London, vol II
125. Hornsby, P.R., Watson, C.L. (1995) J. Mat. Sci. 30: 5347
126. Watson, C.L. (1987) A study of the fire and mechanical properties of polypropylenes filled with sea water magnesium hydroxide, PhD thesis, Brunel University, UK
127. Gilbert, M., Petiraksakul, P. (1996) Proceedings from Loughborough Fillers Symposium II University of Loughborough UK, 17–18th September
128. Hinrichsen, E. (1994) Preparation and characterisation of natural fibre reinforced thermoplastics composites, MPhil thesis, Brunel University, UK
129. Pearson, J.R.A. (1985) In: Mechanics of polymer processing. Elsevier Applied Science, London, chap 18
130. Manas-Zloczower, I., Nir, A., Tadmor, Z. (1985) Polymer Composites 6 (4): 222
131. Yao, C-H., Manas-Zloczower, I. (1996) Polymer Engineering and Science 36 (3): 305
132. Understanding the Banbury (1968) Mixer, Farrel Co., Ansonia USA
133. Min, K., White, J.L. (1985) Rubber Chemistry and Technology 58 (5): 1024
134. Min, K., White, J.L. (1987) Rubber Chemistry and Technology 60 (2): 362
135. Min, K. (1987) Intern. Polym. Process 1 (4): 179
136. Menges, G., Grajewski, F. (1988) Intern. Polym. Process 3 (2): 74
137. Basir, K.B., Freakley, P.K. (1982) Kautch Gummi Kunstst. 35: 205
138. Griffin, H.D. (1984), Inter. Poly. Sci. Technol. 7: 11
139. Kawanishi, K., Yagii, K. (1990) Intern. Polymer Processing 5 (3): 164, 173
140. Cheng, J.J., Manas-Zloczower, I. (1990) Intern. Polymer Processing, 5 (3): 178
141. Nassehi, V., Salemi, R. (1994) Intern. Polymer Processing, 9 (3): 199
142. Nassehi, V., Freakley, P.K. (1991) Intern. Polymer Processing 6: (2) 91
143. Kim, J.K., White, J.L. (1991) Intern. Polymer Processing 6 (2): 103
144. Kawanishi, K., Yagii, K. Obata, Y., Kimura, S. (1991) Intern. Polymer Processing 6 (2): 111
145. Manas-Zloczower, I., Nir A., Tadmor, Z. (1982), Rubber Chemistry and Technology 55: 1250
146. Kearney, M.R. (1991) In: Rauwendaal, C. (ed.) Mixing in polymer processing. Marcel Dekker, New York, chap 4
147. Canedo, E.L., Valsamis, L.N. (1994) Intern. Polymer Processing 9 (3): 225
148. Schenkel, G.P.M. (1988) Intern. Polymer Processing 3 (1): 3
149. Brzoskowski, R., Kumazawa, T., White, J.L. (1991) Intern. Polymer Processing 6 (2): 136
150. Janssen, L.P.B.M., Smith, J.M. (1980) Plastics and Rubber: Processing Sept/Dec: 115
151. Hermann, H., Burkhardt, U. (1980) Plastics and Rubber: Processing Sept/Dec: 101
152. White, J.L. (1990) Twin-screw extrusion – technology and principles. Hanser, Munich
153. Brenner, M.J., Hornsby, P.R. (1989) Plastics & Rubber International 14 (3): 17
154. Szydlowski, W., White, J.L. (1988) Intern. Polymer Processing 2 (3/4): 142
155. Hermann, H. (1988) In: Hensen F. (ed) Plastics extrusion technology. Hanser, Munich, chap 1
156. Ernst, D. (1997) British Plastics & Rubber January 4
157. Berghaus, U., Müller, H. (1995) Plast. Europe (9): 25
158. Allan, P.S., Bevis, M.J., Hornsby, P.R. (1987) Modern Plastics International April: 38
159. Allan, P.S., Bevis, M.J., Gibson, J.R., Hornsby, P.R. (1995) Proceedings from Polymer Process Engineering '95 Conference 11/12 July University of Bradford, UK
160. Bream, C.E., Hinrichsen, E., Hornsby, P.R., Tarverdi, K., Williams, K.S. (1997) Proceedings from Polymer Processing Engineering '97 Conference 1/2 July University of Bradford, UK, Institute of Materials, London

161. Bevis, M.J., Bream, C.E., Hornsby, P.R., Tarverdi, K., Williams, K.S. (1996) Proceedings from Composites Congress '96, Hinckley UK British Plastics Federation
162. Mayer, M. (1988) In: Hensen, F. (ed.) Plastics extrusion technology. Hanser, Munich, chap 17
163. Bowman, J., Rolland, S.M., Coppard, R., Rakowski, R. (1990) Proc. Soc. Plastics Eng. Ann. Tech. Conf., Society of Plastics Engineers Brookfield USA: 69
164. Barker, M.B., Bowman, J., Bevis, M.J. (1983) J. Mat. Sci. 18: 1095
165. Ma, C-Y., White, J.L., Weissert, F.C., Min, K. (1985) Polymer Composites 6 (4): 215
166. Ma, C-Y., White, J.L., Weissert, F.C., Min, K. (1985) J. Non-Newtonian Fluid Mechanics 17: 275
167. Lim, S., White, J.L. (1993) Intern. Polymer Processing 8 (1): 81
168. Lim, S.H., White, J.L. (1990) J. Rheol. 34 (3): 343
169. Suh, C.H., White, J.L. (1996) Poly. Eng. Sci. 32 (11): 1521
170. Suh, C.H., White, J.L. (1996) Poly. Eng. Sci. 36 (17): 2188
171. Medina, E.M.G. (1996) Alignment and management of talc platelets in polypropylene matrix by the application of shear controlled orientation injection moulding (SCORIM) technology. PhD thesis, Brunel University, UK
172. Allan, P.S., Bevis, M.J. (1987) Plastics and Rubber Processing and Applications 7: 3
173. Allan, P.S., Bevis, M.J. (1990) Composites Manufacturing 1 (2) June: 79
174. Fujiyama, M. (1992) Intern. Polymer Processing 7 (2): 165
175. Fujiyama, M. (1992) Intern. Polymer Processing 7 (4): 359
176. Papathanasiou, T.D. (1996) Intern. Polymer Processing 11 (3): 275
177. Sakai, T., Nakamura, K., Morii, A. (1991) Intern. Polymer Processing 6 (1): 26.
178. Bigg, D.M. (1984) Journal of Rheology 28 (5): 501
179. Lee, B-L. (1993) Journal of Vinyl Technology 15 (3): 173
180. Klason, C., Kubát, J. (1984) Proceedings from European Meeting on 'Polymer Processing and Properties' 4:47
181. Bayer, R.K. et al. (1988) J. Mat. Sci. 23: 475
182. Ezqerra, T.A., Bayer, R.K., Balta Calleja, F.J. (1988) J. Mat. Sci. 23: 4121

Received: October 1997

Author Index Volumes 101–139

Author Index Volumes 1-100 see Volume 100

Subject Index

MIX
Papier aus verantwortungsvollen Quellen
Paper from responsible sources
FSC® C105338

If you have any concerns about our products,
you can contact us on
ProductSafety@springernature.com

In case Publisher is established outside the EU,
the EU authorized representative is:
Springer Nature Customer Service Center GmbH
Europaplatz 3, 69115 Heidelberg, Germany

Printed by Libri Plureos GmbH
in Hamburg, Germany